Principles of Biomedical Instrumentation

This accessible yet in-depth textbook describes the step-by-step processes involved in biomedical device design. Integrating microfabrication techniques, sensors and digital signal processing with key clinical applications, it covers:

- the measurement, amplification and digitization of physiological signals, and the removal of interfering signals
- the transmission of signals from implanted sensors through the body, and the issues concerning the powering of these sensors
- networks for transferring sensitive patient data to hospitals for continuous home-monitoring systems
- electrical and biological tests for ensuring patient safety
- the cost–benefit and technological trade-offs involved in device design
- current challenges in biomedical device design.

With dedicated chapters on electrocardiography, digital hearing aids and mobile health, and including numerous end-of-chapter homework problems, online solutions and additional references for extended learning, it is the ideal resource for senior undergraduate students taking courses in biomedical instrumentation and clinical technology.

Andrew G. Webb is Professor and Director of the C. J. Gorter Center for High Field Magnetic Resonance Imaging at the Leiden University Medical Center. He has authored or co-authored several books, including *Introduction to Medical Imaging* (Cambridge University Press, 2010) and *Introduction to Biomedical Imaging* (Wiley, 2002).

CAMBRIDGE TEXTS IN BIOMEDICAL ENGINEERING

Series Editors
W. Mark Saltzman, Yale University
Shu Chien, University of California, San Diego

Series Advisors
Jerry Collins, Alabama A & M University
Robert Malkin, Duke University
Kathy Ferrara, University of California, Davis
Nicholas Peppas, University of Texas, Austin
Roger Kamm, Massachusetts Institute of Technology
Masaaki Sato, Tohoku University, Japan
Christine Schmidt, University of Florida
George Truskey, Duke University
Douglas Lauffenburger, Massachusetts Institute of Technology

Cambridge Texts in Biomedical Engineering provide a forum for high-quality textbooks targeted at undergraduate and graduate courses in biomedical engineering. They cover a broad range of biomedical engineering topics from introductory texts to advanced topics, including biomechanics, physiology, biomedical instrumentation, imaging, signals and systems, cell engineering and bioinformatics, as well as other relevant subjects, with a blending of theory and practice. While aiming primarily at biomedical engineering students, this series is also suitable for courses in broader disciplines in engineering, the life sciences and medicine.

Principles of Biomedical Instrumentation

Andrew G. Webb

Leiden University Medical Center, The Netherlands

CAMBRIDGE
UNIVERSITY PRESS

CAMBRIDGE
UNIVERSITY PRESS

University Printing House, Cambridge CB2 8BS, United Kingdom

One Liberty Plaza, 20th Floor, New York, NY 10006, USA

477 Williamstown Road, Port Melbourne, VIC 3207, Australia

314-321, 3rd Floor, Plot 3, Splendor Forum, Jasola District Centre, New Delhi - 110025, India

79 Anson Road, #06-04/06, Singapore 079906

Cambridge University Press is part of the University of Cambridge.

It furthers the University's mission by disseminating knowledge in the pursuit of
education, learning and research at the highest international levels of excellence.

www.cambridge.org
Information on this title: www.cambridge.org/9781107113138
DOI: 10.1017/9781316286210

First published 2018

A catalogue record for this publication is available from the British Library

Library of Congress Cataloging in Publication data
Names: Webb, Andrew (Andrew G.), author.
Title: Principles of biomedical instrumentation / Andrew G. Webb.
Other titles: Cambridge texts in biomedical engineering.
Description: Cambridge, United Kingdom ; New York, NY, USA : Cambridge University Press,
[2018] | Series: Cambridge texts in biomedical engineering | Includes bibliographical references
and index.
Identifiers: LCCN 2017024373 | ISBN 9781107113138
Subjects: | MESH: Equipment and Supplies | Equipment Design
Classification: LCC R857.B54 | NLM W 26 | DDC 610.28/4–dc23
LC record available at https://lccn.loc.gov/2017024373

ISBN 978-1-107-11313-8 Hardback

Additional resources available at www.cambridge.org/webb-principles

Contents

Preface

The main aim of this textbook is to provide the tools to understand the function and design of different biomedical instruments and devices, and for the reader to be able to use these tools to envision new and improved future technology and designs. Throughout the book the terms medical and biomedical are used interchangeably, and similarly with the terms device and instrument. With several thousand instruments and devices on the market it is clearly impossible to consider more than a handful in detail. Instead, this book concentrates on the following general approach.

(i) What is the clinically relevant measurement that needs to be made?

(ii) What are the pathophysiological mechanisms that give rise to the clinical condition?

(iii) What are the characteristics (e.g. magnitude, frequency, bandwidth) of the signal to be measured? How accurate and reproducible does the measurement have to be? What are the interfering signals that have to be suppressed?

(iv) What are the recommended instrumental specifications for the particular device? How does one design the device to meet these specifications?

(v) How does one test the biocompatibility and electrical safety of such a device, whether it is external or implanted, so that it can operate safely for a number of years?

Traditionally, most of these instruments and devices have been located in a hospital, and patients travel to the clinics for the measurements to be performed by trained personnel. However, this model is changing and nowadays there is an increasing role for what is termed mobile health (m-health), which involves a much greater degree of healthcare being performed by the patients themselves at home. This means that a biomedical device must operate in the patient's home environment without specialized training, as well as transmit the data wirelessly and securely to the physician. This model of **continuous patient monitoring** not only provides a much more complete assessment of the patient's health, but also reduces the number of visits that a patient has to make to the hospital.

The main areas of technological development that have enabled the rapid incorporation of m-health into the healthcare infrastructure are wearable and implantable devices that can transmit data to the physician via mobile phone networks. The integration of micromachined and microfabricated electronics has enabled much smaller high performance devices to be designed. Classic measurement circuits such as the Wheatstone bridge, which used to consist of small lumped element resistors, can now be produced using integrated circuit technology on a submillimetre scale and integrated with amplifiers and filters on a single chip. The role of integrated digital signal processors (DSPs) has become much more important, and has played a key role in fundamental improvements in devices such as hearing aids. The technical challenges of the trade-off between increased signal processing power, miniaturization and battery life are discussed throughout the book.

ORGANIZATION

The skills needed to design and analyze the performance of biomedical instruments come mainly from an engineering background. This book assumes a basic level of understanding of electrical circuits and signal processing, typical of a second-year undergraduate course. No prior knowledge of clinical pathophysiology is assumed, although a basic course in anatomy would be useful.

Chapter 1 outlines the general principles involved in designing a biomedical instrument or device. An outline of the classification schemes used by regulatory bodies and manufacturers is given, along with a general discussion on the steps involved in the design and regulation process. Finally, the various safety standards for biomedical instrumentation are introduced, including testing of hardware, software and user interfaces.

Chapters 2 to 4 cover the basic building blocks of many types of biomedical instrumentation, namely: (i) the transducer/sensor, (ii) the electronic filters and amplifiers, and (iii) the data acquisition system. Analysis of these building blocks involves basic circuit theory and methods. At the end of each of these chapters, examples of the integration of these analysis tools into a real instrument are included. For example, the designs of sequential elements of a pulse oximeter are covered in sections 2.3 (sensor), section 3.6.1 (signal filtering and filtering) and section 4.5.1 (analogue-to-digital conversion).

Chapter 2 begins with a brief introduction to micro-electro-mechanical system (MEMS) devices, which are widely used in modern biomedical devices. Different types of transducers are then described, based on voltage sensors, optical sensors, displacement/pressure sensors, chemical sensors and acoustic sensors. Practical examples are given for each transducer, specifically a biopotential electrode, pulse oximeter, disposable blood pressure monitor, glucose monitor and hearing-aid microphone, respectively.

Chapter 3 concentrates on the design of passive and active filters, as well as the analysis and design of amplification circuits based on operational amplifiers (op-amps). Common higher order filter geometries such as Butterworth, Chebyshev and Sallen–Key are analyzed. Finally, specific amplification and filtering circuits for a pulse oximeter and glucose monitor are shown.

Chapter 4 outlines the different types of analogue-to-digital converter (ADC) architectures that can be used for biomedical instruments, as well as their respective characteristics in terms of resolution and sampling rates. The concepts of oversampling and digital filtering are summarized in the framework of the delta-sigma ADC. The characteristics of different biosignals are described in order to classify them as either deterministic or stochastic: subdivision into different subgroups is also discussed. Finally, different signal processing algorithms including Fourier transformation and time-domain correlation techniques are summarized. Practical examples of correlation analysis in a Swan–Ganz catheter used for cardiac output measurements, as well as short-time Fourier transformation of electroencephalographic signals, are shown.

Chapter 5 deals with the most common measurement using biopotential electrodes, the electrocardiogram (ECG). The design and placement of electrodes, amplification and filtering circuitry, data acquisition system, and integrated signal processing and analysis algorithms are all discussed. Clinical applications that relate the underlying pathophysiology to changes in the ECG signal are outlined, including the most common heart diseases. Finally, the acquisition and analysis of high-frequency ECG signals are described.

Chapter 6 describes the instrumentation associated with a second measurement technique using biopotential electrodes, electroencephalography (EEG). Applications of EEG measurements to diseases such as epilepsy are outlined, as well as its increasing use in monitoring levels of anaesthesia in the operating theatre using the bispectral index. Finally, applications aiding patients with severe physical injuries to communicate and move via brain–computer interfaces are described.

Chapter 7 covers the basics of digital hearing aid design. Different configurations of microphones, and their use in beam-forming techniques to aid speech comprehension in noisy environments, are described. The important role of DSPs in continuous real-time processing within a very small device is discussed in terms of the design trade-offs of performance versus long battery life. Wireless integration of digital hearing aids with external devices, such as mobile phones, is also outlined.

Chapter 8 concentrates on the general topic of m-health and wireless transmission of biomedical data. The role of smartphones and medically related software applications is described, as well as the increasing use of wearable health-monitoring technology. From a medical device point of view the most important developments are the designs of implantable devices, which must be able to connect wirelessly with external transmitters and receivers. Several cardiovascular wireless implanted devices are described, as well as continuous glucose monitors and very new devices such as those for measuring intraocular pressure in patients with glaucoma.

Chapter 9 summarizes the safety issues involved with powered biomedical instruments, both in terms of electrical safety and also biocompatibility. The concepts of electrical macroshock and microshock are introduced, as well as methods of designing equipment to minimize the risks of these events occurring. Equipment for carrying out electrical safety testing is described. The safety of implanted devices is discussed in terms of the biocompatibility of the materials used, as well as their electromagnetic safety. Examples of several biological testing procedures such as cytotoxicity and haemocompatibility are given. Finally, the design of medical devices that can also be used in a magnetic resonance imaging scanner is discussed.

PROBLEMS

The majority of the problems are based specifically around the material in this book, and especially those in Chapters 2 to 4 test knowledge of the circuits, analysis tools and mode of operation of each of the sub-blocks within a biomedical instrument. Some questions may require the writing of relatively simple numerical computer code. Studying instrumentation design is not only a question of understanding the current state of the art but also constitutes a platform to envision alternative approaches, and to speculate on which new technologies and improvements may be incorporated in the future. Therefore, especially in Chapters 5 to 9, there are a number of problems that require the use of external sources in order to investigate devices that are not covered specifically in this book. For example, with the knowledge gained from Chapters 2 to 4 how would one design a new technique for harvesting the body's internal energy for powering an implanted sensor (Problem 8.4). In a similar vein, investigating cases of device failure is a good way to see how apparently watertight safety procedures can, in fact, fail and to consider the implications of these failures for future designs (Problem 1.3).

REFERENCE BOOKS

There is a large number of books covering different aspects of the principles and design of biomedical instrumentation and devices. The unofficial 'bible' of general biomedical instrumentation remains *The Encyclopedia of Medical Devices and Instrumentation* edited by J. G. Webster and published by Wiley-Blackwell in 2004, as well as the distilled classic textbook *Medical Instrumentation: Application and Design*, 4th edition, J. G. Webster, John Wiley & Sons, 2010.

In terms of books that concentrate on one specific aspect of design, the list below represents a partial list that have proved very useful in preparing this current volume.

Chapter 1 Yock, P. G., Zenios, S., Makower, J. *et al.* (eds) *Biodesign: The Process of Innovating Medical Technologies.* Cambridge University Press; 2015.

Chapter 2 Jones D. P. and Watson. J. *Biomedical Sensors* (Sensor Technology Series). Momentum Press; 2010.

Chapter 3 Huijsing, J. *Operational Amplifiers: Theory and Design, 3rd edn.* Springer; 2016.

Chapter 4 Pelgrom, M. J. *Analog-to-Digital Conversion.* Springer; 2013.

Chapter 5 Crawford, J. & Doherty, L. *Practical Aspects of ECG Recording.* M&K Update Ltd; 2012.

Chapter 6 Libenson, M. H. *Practical Approach to Electroencephalography.* Saunders; 2009.

Chapter 7 Popelka, G. R., Moore, B. C. J., Fay, R. R. & Popper, A. N. *Hearing Aids.* Springer; 2016.

Chapter 8 Salvo, P. & Hernandez-Silveira, M. (eds) *Wireless Medical Systems and Algorithms: Design and Applications (Devices, Circuits, and Systems).* CRC Press; 2016.

Chapter 9 Gad, S. C. & McCord, M G. *Safety Evaluation in the Development of Medical Devices and Combination Products, 3rd edn.* CRC Press; 2008.

Abbreviations

AAMI	Association for the Advancement of Medical Instrumentation
AC	alternating current
ADC	analogue-to-digital converter
ADHF	acute decompensated heart failure
AF	atrial fibrillation
AGC	automatic gain control
ANSI	American National Standards Institute
ASIC	application-specific integrated circuit
ASK	amplitude-shift keying
ATP	adenosine triphosphate
AV	atrioventricular
AVC	automatic volume control
BCI	brain–computer interface
BcSEF	burst-compensated spectral edge frequency
BF	body floating
BILL	bass increase at low level
BIS	bispectral index
BiV	biventricular
BSR	burst-suppression ratio
BTE	behind the ear
BW	bandwidth
CAD	computer-aided design
CE	Conformité Européene
CF	cardiac floating
CFR	Code of Federal Regulations
CGM	continuous glucose monitor
CIC	completely in canal
CIED	cardiovascular implantable electronic device
CMRR	common-mode rejection ratio
CRT	cardiac resynchronization therapy
CSF	cerebrospinal fluid
DAC	digital-to-analogue converter
dB	decibel

DC	direct current
DDD	dual mode, dual chamber, dual sensing
DF	directivity factor
DI	directivity index
DNL	differential non-linearity
DRL	driven right leg
DSP	digital signal processing
EC	European Commission
ECG	electrocardiogram
EDR	electrodermal response
EEG	electroencephalogram
EMG	electromyogram
EMI	electromagnetic interference
EOG	electrooculogram
EPROM	erasable programmable read-only memory
EPSP	excitatory postsynaptic potential
ESI	electric source imaging
FCC	Federal Communications Commission
FDA	Food and Drug Administration
FES	functional electrical stimulation
FET	field-effect transistor
FFT	fast Fourier transform
FIR	finite impulse response
FM	frequency modulation
FSK	frequency shift keying
FT	Fourier transform
GBWP	gain bandwidth product
GDH	glucose-1-dehydrogenase
GFCI	ground-fault current interruptor
GHK	Goldman–Hodgkin–Katz
GOx	glucose oxidase
GPS	global positioning system
GSM	global systems for mobile
GSR	galvanic skin response
HPF	high-pass filter
HR	heart rate
ICA	independent component analysis
ICD	implantable cardioverter–defibrillator
ICNIRP	International Commission on Non-Ionizing Radiation Protection
ICU	intensive care unit

IDE	investigational device
IEC	International Electrotechnical Commission
IEEE	Institute of Electrical and Electronic Engineers
IHM	implantable haemodynamic monitor
IIR	infinite impulse response
ILR	implantable loop recorder
INL	integrated non-linearity
IOL	intraocular lens
IOP	intraocular pressure
IPAP	implanted pulmonary artery pressure
IPSP	inhibitory postsynaptic potential
ISM	industrial, scientific and medical
ISO	International Organization for Standardization
ITC	in the canal
ITE	in the ear
JFET	junction field-effect transistor
KEMAR	Knowles Electronics Manikin for Acoustic Research
LA	left arm
LBBB	left bundle branch block
LED	light-emitting diode
LIM	line-isolation monitor
LL	left leg
LPF	low-pass filter
LSB	least significant bit
LVDT	linear voltage differential transformer
MD	medical device
MDRS	medical device radiocommunication service
MEMS	micro-electro-mechanical systems
MI	myocardial infarct
MICS	medical implant communication service
MRI	magnetic resonance imaging
MSB	most significant bit
MUX	multiplex
NB	national body
NC	normal conditions
NEC	National Electrical Code
NF	noise figure
NFMI	near-field magnetic induction
NIDCD	National Institute on Deafness and Other Communication Disorders
NSTEMI	non-ST-segment elevated myocardial infarction

OR	operating room
PA	pulmonary artery
PAC	pulmonary artery catheter
PCA	principal component analysis
PDF	probability density function
PDMS	polydimethylsiloxane
PIFA	planar inverted F-antenna
PILL	programmable increase at low level
PIPO	parallel input, parallel output
PISO	parallel input, serial output
PMA	pre-market approval
P-N	positive–negative
PSK	phase shift keying
PSP	postsynaptic potential
PTFE	polytetrafluoroethylene
PVARP	post-ventricular atrial refractory period
PVDF	polyvinyldifluoride
PWM	pulse-width modulation
QSR	quality-system regulation
RA	right arm
RAM	random access memory
RAZ	reduced amplitude zone
RBC	red blood cell
RCD	residual-current device
RF	radiofrequency
RLD	right leg drive
ROC	receiver operator characteristic
ROM	read-only memory
RPM	remote patient monitoring
RV	right ventricle
S/H	sample and hold
SA	sinoatrial
SAR	specific absorption rate
SAR ADC	successive approximation register ADC
SC	single condition
SCP	slow cortical potential
SELV	separated extra-low voltage
SFC	single-fault condition
SFDR	spurious-free dynamic range
SINAD	signal-to-noise and distortion ratio

SIP	signal input
SIPO	serial input, parallel output
SISO	serial input, serial output
SNR	signal-to-noise ratio
SOP	signal output
SPL	sound pressure level
SpO_2	saturation of peripheral oxygen
SSVEP	steady state visual evoked potential
SVI	single ventricular
TARP	total atrial refractory period
THD	total harmonic distortion
TILL	treble increase at low level
UWB	ultrawide band
VEP	visual evoked potential
VLP	ventricular late potential
VOP	venous occlusion plethysmography
WCT	Wilson's central terminal
WDRC	wide dynamic range compression
WMTS	wireless medical telemetry system

1 Biomedical Instrumentation and Devices

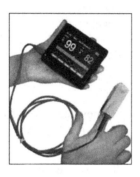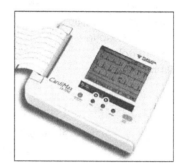

1.1 Classification of Biomedical Instruments and Devices

Worldwide the medical instrumentation and device industry is worth more than 100 billion US dollars annually. A number of multinational companies, including Boston Scientific, Medtronic, Abbot Medical Devices, Johnson & Johnson and Novo Nordisk, have a major focus on the development, sales and distribution of several broad classes of medical devices. In the United States, the five largest areas of medical device revenue are orthopaedics, ophthalmology, cardiology, audiology and surgery, each with revenues of about 20 billion US dollars.

Table 1.1 Device classification panels: upper tier (www.fda.gov/MedicalDevices/DeviceRegulationandGuidance)

Medical specialty	Regulation citation	Medical specialty	Regulation citation
Anaesthesiology	Part 868	Microbiology	Part 866
Cardiovascular	Part 870	Neurology	Part 882
Chemistry	Part 862	Obstetrical and gynaecological	Part 884
Dental	Part 872	Ophthalmic	Part 886
Ear, nose and throat	Part 874	Orthopaedic	Part 888
Gastroenterology and urology	Part 876	Pathology	Part 864
General and plastic surgery	Part 878	Physical medicine	Part 890
General hospital	Part 880	Radiology	Part 892
Haematology	Part 864	Toxicology	Part 862
Immunology	Part 866		

Hundreds of smaller companies, including new start-ups, concentrate on more specialized parts of the market. A search of the US-based Food and Drug Administration (FDA) Code of Federal Regulations (CFR) in 2014 showed that over 1700 distinct types of (bio)medical devices and instruments are listed. The FDA defines a medical device as:

> An instrument, apparatus, implement, machine, contrivance, implant, in vitro reagent, or other similar or related article, including a component part, or accessory which is [...] intended for use in the diagnosis of disease or other conditions, or in the cure, mitigation, treatment or prevention of disease, in man or other animals, or intended to affect the structure or any function of the body of man or other animals, and which does not achieve any of its primary intended purposes through chemical action within or on the body of man or other animals and which is not dependent upon being metabolized for the achievement of any of its primary intended purposes.

Medical devices are classified in many tiers and sub-tiers. Table 1.1 shows the upper classification tier, which is based on medical specialty.

The next level of classification is illustrated in Table 1.2 using Part 870 on cardiovascular devices as an example. Each of these generic types of devices is assigned to one of three regulatory classes based on the level of control necessary to assure the safety and effectiveness of the device. Class I devices have the lowest risk, Class II intermediate risk and Class III are those with the highest risk. Devices for the European market have similar classes. The class of the medical device determines the type of premarketing submission/application required for regulatory approval: this process is considered in more detail in section 1.3. Each of the elements in Table 1.2 has a small section that describes in very general terms what the device does and its classification in terms of performance standards. As an example:

Table 1.2 Device classification panels: Subpart B cardiovascular diagnostic devices

Device	Section	Device	Section
Arrhythmia detector and alarm (including ST-segment measurement and alarm)	870.1025	Catheter guide wire	870.1330
Blood pressure alarm	870.1100	Catheter introducer	870.1340
Blood pressure computer	870.1110	Catheter balloon repair kit	870.1350
Blood pressure cuff	870.1120	Trace microsphere	870.1360
Non-invasive blood pressure measurement system	870.1130	Catheter tip occluder	870.1370
Venous blood pressure manometer	870.1140	Catheter stylet	870.1380
Diagnostic intravascular catheter	870.1200	Trocar	870.1390
Continuous flush catheter	870.1210	Programmable diagnostic computer	870.1425
Electrode recording catheter or electrode recording probe	870.1220	Single-function, preprogrammed diagnostic computer	870.1435
Fibreoptic oximeter catheter	870.1230	Densitometer	870.1450
Flow-directed catheter	870.1240	Angiographic injector and syringe	870.1650
Percutaneous catheter	870.1250	Indicator injector	870.1660
Intracavitary phonocatheter system	870.1270	Syringe actuator for an injector	870.1670
Steerable catheter	870.1280	External programmable pacemaker pulse generator	870.1750
Steerable catheter control system	870.1290	Withdrawal-infusion pump	870.1800
Catheter cannula	870.1300	Stethoscope	870.1875
Vessel dilator for percutaneous catheterization	870.1310	Thermodilution probe	870.1915

Sec. 870.1130 Non-invasive blood pressure measurement system.
    (a) Identification. A non-invasive blood pressure measurement system is a device that provides a signal from which systolic, diastolic, mean, or any combination of the three pressures can be derived through the use of transducers placed on the surface of the body.
    (b) Classification. Class II (performance standards).

1.2   Outline of the Design Process: From Concept to Clinical Device

Figure 1.1 gives a schematic of the process of producing a new medical instrument or device. Each of the steps is discussed briefly in the next sections, with the exception of reimbursement assignment, which although an obviously important topic, varies widely by country and political philosophy. With the ever-increasing costs of healthcare there is a strong recent trend towards what is termed **'value-based' medicine**, i.e. recognizing that there is a trade-off between improved

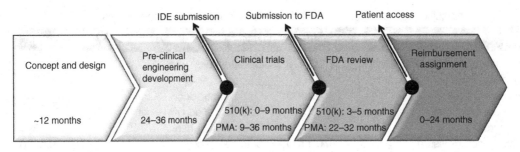

Figure 1.1 General outline of the steps involved in the design, engineering and safety testing, submission process and, ultimately, financial planning for a new medical instrument or device.

healthcare and its availability to the public due to the associated increased costs. Overall, the aim is high-quality care at an affordable cost. As an example of why value-based medicine has become an important concept, consider that the share of the total economy of the United States taken up by healthcare has more than doubled since the 1970s, with a total budget of over 20% of gross domestic product in 2016. However, a recent report stated that the "USA stands out for not getting good value for its healthcare dollars". In the past, new products (both medical drugs and devices) were designed, approved and integrated into medical care even if their costs far outweighed slight improvements in performance and clinical diagnosis. The new paradigm is to calculate the 'value' of a new product, with value defined as the degree of improvement relative to its increased cost (which also includes associated costs in terms of parameters such as the need for highly trained operators and additional training). The implication of this new approach is that medical device manufacturers should aim new products at healthcare areas that are outliers in terms of their low cost-effectiveness, and so can be improved the most, rather than on incremental increases in the performance of devices that already provide good value for money.

1.2.1   Engineering Design

One of the most common comments about the role of engineering in medicine is that engineers are very good at discovering a new concept or technology, and then try desperately to find a clinical application for this technology! This approach, of course, runs counter to every fundamental concept in design engineering: first define the problem and the goal, and *then* design the solution. During the invention phase, it is critically important to have an appreciation of the anatomy and disease pathophysiology.

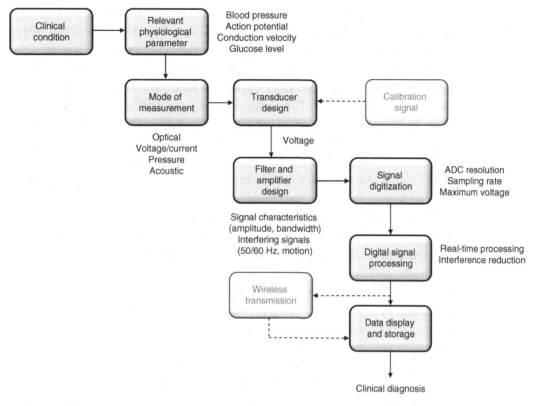

Figure 1.2 Generalized block diagram of the individual modules of a biomedical instrument. Grey text and dotted lines represent features that are only present in certain instruments.

Figure 1.2 shows a general block diagram of the individual components of a biomedical instrument, with examples of different measurements, sensors, filters and data acquisition systems. Often the desired measurement is based inside the body, but ideally the measurement is non-invasive where possible: this means that the signal has to be measured indirectly. How can this signal be measured with the highest fidelity? This involves removing interfering signals from external sources, including other physiological processes such as motion, which may produce signals many times larger than the ones we are trying to measure. Many of the biosignals have a very low magnitude: biopotentials are in the microvolt (electroencephalography) to millivolt (electrocardiography) range, internal pressures are on the orders of kilopascals (blood pressure sensors) and biocurrents lie in the microamp to milliamp range (glucose sensors). The frequencies of the biosignals are also rather low, generally in the range DC to hundreds of hertz (although acoustic signals for hearing aids are in the DC–16 kHz range), meaning that the hardware can be designed for low-frequency filtering and amplification (operational amplifiers).

However, when implanted sensors are needed, power and signal transmission through the body requires much higher frequencies (typically around 400 MHz) and so high-frequency circuits must be integrated into these types of devices.

As an example of the general concepts involved in medical instrumentation design, consider one of the most commonly used machines in a hospital, the electrocardiogram (ECG). This instrument measures how healthy the heart is in terms of its fundamental purpose of pumping blood around the body. The pumping action is caused by action potentials occurring in the pacemaker node of the heart and this electrical activity spreading throughout the heart with well-defined spatial and temporal characteristics, in turn causing the heart as a whole to expand and contract to produce the desired pumping function.

Different pathophysiologies of the heart cause different problems with the conduction path of the action potentials through the heart (**clinical condition**). The fundamental measurements that reflect physiological changes are the action potentials in different areas of the heart (**relevant physiological parameter**). However, the measurement must be performed outside of the body. By analyzing the propagation of ionic currents, produced by the action potentials, through the body, electrical activity in the heart can be detected indirectly on the surface of the body using a number of electrodes placed at different locations (**mode of measurement**). These electrodes transform/transduce the ionic currents into an electrical voltage (**transducer design**). In addition to the desired ECG voltage, there are many electrical interference signals that are also detected by the electrodes, and these interferences can be filtered out knowing the respective frequency ranges over which the ECG signals and interference signals occur (**filter design**). In order to digitize the filtered ECG signal with a high degree of accuracy, signal amplification is needed to use the full dynamic range of the analogue-to-digital converter (**amplifier design, signal digitization**). After digitization, further filtering of the signal can be performed using very sharp digital filters to remove artefacts such as baseline wander or any remaining 50/60 Hz noise (**digital signal processing**). Finally, by analyzing the shape and general characteristics of the ECG voltage waveform in each of the electrodes (**data display**), it is possible to detect abnormalities and to trace these abnormalities back to specific malfunctions of the heart.

Table 1.3 outlines the characteristics of the ECG signal, as well as those of different interference signals, which can actually have a much higher amplitude than the ECG signal itself.

Figure 1.3 shows a block diagram of the individual components of the ECG data acquisition system, which are covered in much more detail in Chapter 5.

Figure 1.4 shows an example of a condition, atrial fibrillation, which produces a clear alteration in the ECG signal. Atrial fibrillation can be caused by

Table 1.3 Design criteria for the electrocardiograph

Physiological metric	Cardiac action potential (voltage)
Indirect measurement device	Electrode (placed on the skin)
Mode of action	Converts ionic current into a voltage
Size of detected signal	1–10 mV
Frequency of detected signal	1–50 Hz
Interfering signals	Electrode half-cell potential; coupling to power lines; breathing; muscle motion; blood flow
Size of interfering signals	1.5 V (power lines); 300 mV (half-cell potential); ~mV (muscle motion)
Frequency of interfering signals	50/60 Hz (power lines); DC (half-cell potential); ~10–50 Hz (muscle motion); ~0.5 Hz (breathing)
Required time resolution	One measurement every 200 ms
Required accuracy	±1 mV
Required dynamic range	0–100 mV

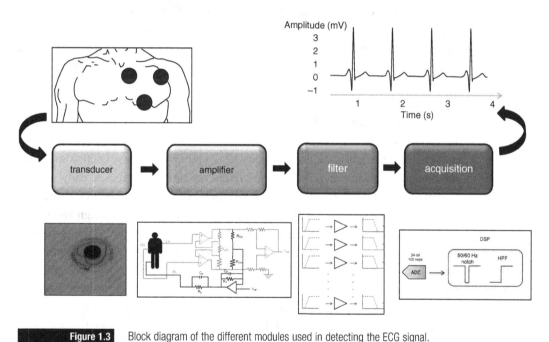

Figure 1.3 Block diagram of the different modules used in detecting the ECG signal.

a number of different conditions including heart failure, coronary artery disease, disease of the heart valves, diabetes and hypertension. Although the waveform shown in Figure 1.4 is highly characteristic of atrial fibrillation, the final diagnosis of the particular condition is usually made by combining the ECG with other diagnostic techniques such as ultrasound and computed tomography.

Figure 1.4

(a) Example of a normal ECG signal in a healthy person. (b) Example of atrial fibrillation.

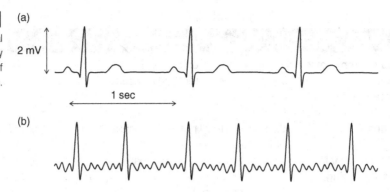

1.3 Regulation of Biomedical Instrumentation and Devices

Regulatory bodies such as the FDA and European Commission (EC) do not set specific regulations per se, but rather rely on standards that are produced by working groups such as the International Organization for Standardization (ISO) and the International Electrotechnical Commission (IEC), as well as country-specific organizations such as the American National Standards Institute (ANSI) and the European Committee for Standardization (CEN).

Initially, a company in the United States has to make a best-guess selection of the appropriate class (I, II or III) of their device. Class I devices do not need to undergo clinical trials or biocompatibility tests, since by the Class I definition there is sufficient evidence of safe operation based on similar devices already on the market. There are still some requirements, such as adhering to the FDA's quality systems regulation (QSR), which represents guidelines for safe design and manufacturing. Class II devices are usually taken to market via the 510(k) process. The particular medical device must be shown to be substantially equivalent to an existing design, so that the FDA can compare it. Equivalence can usually be assessed using bench tests and animal testing without the need for human trials: only about 10% of annual 510(k) submissions require clinical data. A second route for Class II devices is the de novo 510(k) path, which is for devices that do not have the risks associated with Class III devices, but for which no substantial prior information exists or similar devices are not yet on the market. These require a higher level of proof of efficiency than the standard 510(k) but less than for the pre-market approval (PMA) required for Class III devices. The 510(k) submissions are standardly reviewed on the timescale of a few months, but the amount of paperwork is substantial, running to several hundreds of pages. Class III devices require PMA regulatory approval, and represent devices that have the highest potential risk to patients or have significantly different technology than those that already exist in the target application field. Typically, large multi-centre randomized clinical trials are required for these types of devices. If a device, such as a coronary stent, is defined to be 'life-sustaining', then this type of Class III device requires a PMA, even though the new stent may be very similar to those already on the market [1].

One obvious question is how can one actually perform clinical trials on devices that are only officially approved after a successful clinical trial? This process requires an investigational device exemption (IDE), which represents official permission to begin such a trial. For low-risk devices, the local institutional review board or medical ethics committee at the hospital or laboratory where testing is to be performed, can give such approval. For devices that ultimately will require a PMA, clearance must be given by the FDA. The IDE does not allow a company to market the device, merely to perform the clinical trials required to obtain a PMA. However, in the United States companies can charge for investigational devices. The requirements for receiving an IDE are usually extensive biological and animal testing (covered in detail in Chapter 9).

The description above refers explicitly to the situation in the United States, but there are broadly equivalent regulatory standards in the European Union (EU), with some important differences [2, 3]. There are three different European directives: (i) implantable devices are regulated under directive 90/385/EC; (ii) most other devices are regulated under directive 93/42/EC; and (iii) in vitro diagnostic devices (i.e. used on substances produced by the body) are regulated under 98/79/EC [4]. In the EU, every marketed medical device must carry a Conformité Européenne (CE) mark indicating that it conforms to relevant directives set forth in the EC Medical Device Directives. A device with a CE mark can be marketed in any EU member state. Devices are classified as low risk (Class I), moderate risk (Classes IIa and IIb) and high risk (Class III). Medical devices that are non-implantable and considered low risk are 'self-marked', meaning that the manufacturer itself simply certifies compliance and applies a CE mark. High-risk devices must undergo a more extensive outside review by a 'Notified Body' (NB) within that country, which is authorized by that country's Competent Authority, or health agency, to assess and assure conformity with requirements of the relevant EC directive. One of the fundamental differences between the regulatory systems in the United States and Europe is that before approval can be granted for a medical device in the United States, it must not only be shown to be safe, but efficacious. In contrast, medical devices approved in Europe need only demonstrate safety and performance, i.e. they perform as designed and that potential benefits outweigh potential risks: they are not required to demonstrate clinical efficacy.

1.4 Safety of Biomedical Instrumentation and Devices

The safety of a biomedical instrument or device refers to three different facets of the equipment: the hardware, the software and the user interface. The general principles are that in hardware, two *independent failures* should not harm the patient; software is designed such that the chances of harm arising from *inevitable* bugs are acceptably low; and the design of the user interface concentrates on making the man–machine interface as safe as possible. The increasing use of mobile health (m-health) applications has resulted in new FDA guidelines. The FDA has defined a mobile app to constitute a medical device 'if a mobile app is intended for use in performing a medical function (i.e. for diagnosis of disease or other conditions, or the cure, mitigation, treatment, or prevention of disease), it is a medical device, regardless of the platform on which it is run'. Mobile health technology is classified into Classes I, II or III in the same way as physical medical devices. For example, a mobile app that controls a glucose monitor, and stores and transmits the data wirelessly to the physician, is subject to exactly the same regulations as the glucose monitor itself.

Despite all of the safety requirements in place, it is estimated that roughly 1.5% of 501(k) predicate devices have to be recalled. There are also several instances of major recalls involving tens or hundreds of patient deaths (see Problems).

1.4.1 ISO and IEC Standards

As mentioned earlier, regulatory bodies do not design specific safety tests that medical devices must pass, but instead rely on standards that have been devised by independent testing agencies. For example, ISO 14971:2007 specifies the procedure by which medical device manufacturers can identify potential hazards in order to estimate and evaluate the associated risks. Methods must be provided to control these risks, and to monitor the effectiveness of the control methods. Figure 1.5 shows a schematic of the methods for categorizing risks [5].

Provided below is a brief summary of key elements of ISO 14971:2007 with sections in italics (not present in the original document) outlining the essential concepts.

> The requirements contained in this International Standard provide manufacturers with a framework within which *experience, insight and judgment* are applied systematically to manage the risks associated with the use of medical devices. This International Standard was developed specifically for medical device/system manufacturers using established principles of risk management [. . .] This International Standard deals with processes for *managing*

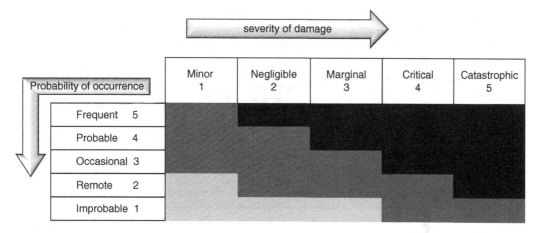

Figure 1.5 The ISO 14971 standard for product risk management outlines methods for categorizing risks according to the probability of their occurring, and the severity of the particular failure. The darker the box the less acceptable the risk.

risks, primarily to the patient, but also to the operator, other persons, other equipment and the environment [. . .]

It is accepted that the concept of risk has two components:

(a) the *probability* of occurrence of harm;

(b) the *consequences* of that harm, that is, how severe it might be.

All stakeholders need to understand that *the use of a medical device entails some degree of risk* [. . .] The decision to use a medical device in the context of a particular clinical procedure requires the *residual risks to be balanced against the anticipated benefits* of the procedure. Such judgments should take into account the *intended use*, performance and risks associated with the medical device, as well as *the risks and benefits* associated with the clinical procedure or the circumstances of use [. . .] As one of the stakeholders, the manufacturer makes judgments relating to safety of a medical device, including the *acceptability of risks, taking into account the generally accepted state of the art*, in order to determine the suitability of a medical device to be placed on the market for its intended use. This International Standard specifies a process through which the manufacturer of a medical device can identify hazards associated with a medical device, estimate and evaluate the risks associated with these hazards, control these risks, and monitor the effectiveness of that control.

Figure 1.6 illustrates the steps involved in all stages of risk management according to ISO 14971:2007.

1.4.1.1 Hardware

The main standard for medical device safety is IEC 60601–1 – Medical Electrical Equipment – Part 1: General Requirements for Basic Safety and Essential Performance. One of the fundamental principles of IEC 60601–1 is that a medical device must be safe in the case of a *single fault*, i.e. the failure of one safety system.

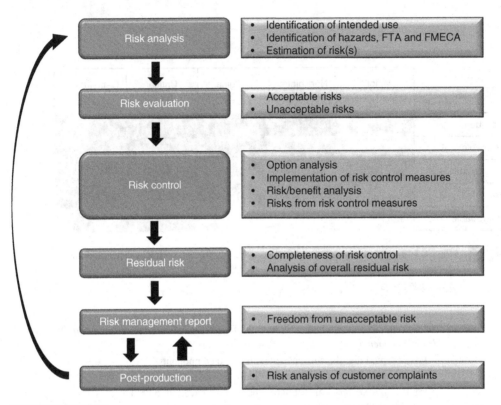

Figure 1.6   Life-cycle approach to risk management adapted from ISO 14971. Acronyms used are failure modes, effects and criticality analysis (FMECA) and fault tree analysis (FTA).

Therefore the starting point for designing the hardware is to have a back-up for every possible 'first-line' failure. Of course, this safety system has its own possibility of failure, and so the next design stage is to perform a risk assessment to determine whether a back-up to the back-up is necessary. The reliability of a safety system can be improved either by periodic checking (either by the operator or via internal system self-checks) or by using two redundant (independent) safety systems. A second fundamental requirement of IEC 60601–1 is that a combination of two independent failures should not be life threatening. If the first failure is obvious to the operator, he/she can stop the equipment immediately, and the equipment can be taken out of commission and repaired. However, if the first failure is not immediately obvious or cannot be detected, it must be assumed that a second failure will occur sometime later. The equipment must be designed so that a combination of the first and second failures cannot cause a hazard.

Hardware failures can in general be characterized as either **systematic** or **random**. Systematic failures are basically built-in design flaws. Examples include errors

Figure 1.7

Illustration of the V-model
commonly used for
system safety checking.

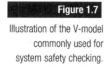

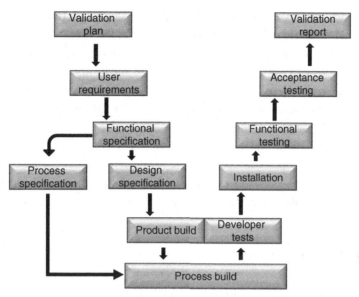

in the layout of a printed circuit board, or electrical components being used outside their specifications or encountering unanticipated environmental conditions such as high humidity levels. Random errors, on the other hand, occur even though the design is fundamentally correct and production is flawless. Random hardware errors cannot be predicted on an individual basis, but only be described statistically. The general approach to controlling random errors is by incorporating redundant features, or by adding self-testing or a safety system that reacts in the event of a random failure.

1.4.1.2 Software and User Interfaces

Software for medical devices has its own regulatory standard, IEC 62304, 'Medical device software – software life cycle processes' that specifies requirements for the development of both standalone medical software and also software used in medical devices [6]. IEC 62304 uses the risk management process outlined in ISO 14971, with additions related to hazards such as incorrect information provided by the software resulting in an inappropriate treatment being prescribed. The primary principle behind software verification is to describe the exact function the software is supposed to perform, and then to devise a very specific test to verify that the software works as designed.

Software safety analysis often employs a V-model (which is also used in many other system designs), shown in Figure 1.7. This model describes the decomposition of software requirements into three levels (the descending arm of the V), and the testing and validation from the smallest software unit that can be tested fully (lowest level) through to the highest system level (the ascending arm of the V). A software risk model defines three different levels of safety. Level A software is not harmful if it fails; Level C software can injure or kill someone; and Level B represents an intermediate level. Most medically related software is assigned to Level C. Level A software requires a system test; Level B requires checks of the individual software modules; and Level C software requires tests for each subset of software code at the unit level.

With regards to user interfaces, standards for human–computer interfaces are much less developed than for hardware or software. Indeed, the major recommendation of IEC 62366 Annex D 1.4 simply requires designers to analyze how users will interact with the device and to implement risk-mitigating measures to avoid erroneous usage.

1.4.2 Biological Testing

The biocompatibility of different devices, particularly implanted devices, is covered in detail in Chapter 9. As a brief summary, biocompatibility cannot be determined or defined by any one single test and therefore as many appropriate biocompatibility parameters as possible should be tested. One key aspect of biocompatibility testing is that the tests should be designed to assess potential adverse effects under **actual use conditions** or conditions as close as possible to the actual use conditions. The exact battery of tests required to ensure biocompatibility depends on the nature and duration of any physical contact between the particular device and the human body. Some devices, such as those used in cardiovascular implants covered in Chapter 8, are designed to be implanted in the patient for many years. In these cases, the biocompatibility testing procedure needs to show that the implant does not have any adverse effects over such a long period of time. The ISO-10993 document, again expanded upon in Chapter 9, provides detailed guidelines on biocompatibility testing.

1.5 Evaluation of a New Device

After a new medical device has been designed and has undergone basic safety testing, its performance should be compared to other devices that are already on the market. Although the new device might have a better performance, the key question is whether this actually translates to a higher diagnostic value, particularly with regards to the value-based criteria outlined earlier in this chapter.

Quantitatively, one can assess the performance of a device using measures such as its accuracy, resolution, precision and reproducibility, and then relate these measures to its sensitivity and specificity. The **accuracy** of a device is defined as the difference between the measured value and the true value measured by the 'gold-standard' system. An example is an implanted glucose meter, whose readings are compared to those from sophisticated laboratory-based glucose meters. The accuracy is usually expressed as a percentage, and is assumed to be equally distributed above and below the true reading, e.g. a glucose reading of 95 mg/dL +/–5%. The accuracy may also depend upon the absolute value of the reading, for example it may be more accurate for higher readings and less accurate for lower readings. The **resolution** of an instrument is defined as the smallest differential value that can be measured: for example, an analogue-to-digital converter, as its name suggests, converts an analogue signal to a digital one. The digital readings can only have specific values, e.g. 2 mV, 3 mV or 4 mV, and the difference between these discrete values is the instrument resolution. The resolution is closely related to the **precision** of the instrument, which is a measure of the lowest difference between measurements that can be detected. The **reproducibility** is an indication of the difference between successive measurements taken under identical measuring conditions.

New instruments and devices can have higher accuracy, resolution and reproducibility than existing devices, and yet not make an actual difference to clinical diagnosis or treatment monitoring. For example, if a new implantable glucose sensor is 1% more accurate in terms of measuring plasma glucose levels, this will not result in any difference in the effective operation of the device. There are several quantitative measures by which the clinical effect of improvements in device performance can be assessed. These measures can be very specific to the type of measurement being performed: for example, the widely used Parkes error grid for glucose measurements, discussed in detail in section 8.4.2. However, there are also several general tools such as the receiver operating characteristic (ROC) curve, which can be considered as a binary classification in which the outcomes are either positive or negative and the predictions are either true or false. Therefore there are four possibilities: true positive (TP), false positive (FP), true negative (TN) and false negative (FN). For example, if a new continuous glucose monitoring system suggests that, based on the trend of glucose levels over time, a patient will become hypoglycaemic, TP indicates that hypoglycaemia did indeed occur, whereas FP corresponds to the case where it did not occur. Conversely, if the monitor suggests that hypoglycaemia will not occur and this was the correct prognosis, then this corresponds to a TN; if hypoglycaemia in fact occurred then this was a FN. The ROC plot is a graph of the TP rate on the vertical axis versus the FP rate on the horizontal axis, as shown in Figure 1.8. A straight line at 45° to the

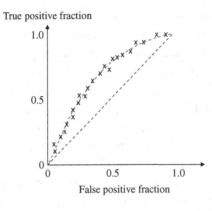

Figure 1.8

Example of a receiver operating characteristic curve (grey dashed line) for a medical instrument. The dotted line represents the performance of an instrument that has no prognostic value.

axis corresponds to essentially a random measurement with no prognostic value. The more the ROC curve lies above this line, the better the device. In this way, a new device can be compared with an existing one. Two very commonly quoted parameters for a device are the sensitivity and specificity. The **sensitivity** of a device is defined as the number of true positives divided by the sum of the true positives and false negatives, and the **specificity** is the number of true negatives divided by the sum of the number of true negatives and false positives.

PROBLEMS

1.1 Choose one medical device in the FDA list (Tables 1.1 and 1.2) and look up the history in terms of the original idea, patenting the original idea, the FDA approval process and any recent technical developments in the device or instrument.

1.2 Look up the history of the most famous failed devices, for example the Guidant LLC implantable cardioverter–defibrillator; Medtronic Sprint Fidelis implantable cardioverter-defibrillator; or St. Jude Medical Riata ST defibrillator leads. What were the major design flaws? How were these detected? To what extent was faulty testing performed, and what has been improved in the testing procedure to deal with these problems?

1.3 Go to the FDA recalls site: www.fda.gov/MedicalDevices/Safety/ListofRecalls. Choose one recall from each class, summarize the problem, outline the solution and discuss the implications.

1.4 Can a random error be introduced by a systemic software error? Give one or more examples.

1.5 Draw the ideal ROC curve for a device. Can a device have a ROC curve that lies below the dotted line in Figure 1.8? If so, suggest one situation involving a medical device that could produce such a line.

REFERENCES

[1] Resnic, F. S & Normand, S. L. T. Postmarketing surveillance of medical devices: filling in the gaps. *New Engl J Med* 2012; **366**(10):875–7.

[2] Kramer, D. B., Xu, S. & Kesselheim, A. S. Regulation of medical devices in the United States and European Union. *New Engl J Med* 2012; **366**(9):848–55.

[3] Van Norman, G. A. Drugs and devices: comparison of European and U.S. approval processes. *JACC Basic Transl Sci* 2016; **1**:399–412.

[4] Campbell, B. Regulation and safe adoption of new medical devices and procedures. *Brit Med Bull* 2013; **107**(1):5–18.

[5] Flood, D., McCaffery, F., Casey, V., McKeever, R. & Rust, P. A roadmap to ISO 14971 implementation. *J Softw Evol Proc* 2015; **27**(5):319–36.

[6] Rust, P. Flood, D. & McCaffery, F. Creation of an IEC 62304 compliant software development plan. *J Softw Evol Proc* 2016; **28**(11):1005–110.

2 Sensors and Transducers

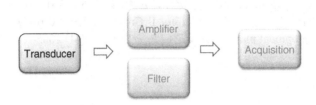

Introduction

The term **sensor** refers to a device that is used to detect a signal. A sensor often forms part of a 'transducer', which converts one type of input signal (e.g. ionic current, mechanical vibration, absorbed light or chemical reaction) into another form of output signal. In most biomedical devices the output signal is a voltage, which can be either DC or AC. This signal voltage is then filtered and amplified before being digitized by an analogue-to-digital converter (ADC), after which it is stored/displayed as digital data. Additional signal processing can then be performed on the digitized data as required.

In this chapter, five different types of sensors are considered: voltage, optical, mechanical, chemical and acoustic. For each of these sensors the following topics are covered:

(i) the clinical need for such a measurement
(ii) the underlying physiological basis of the detected signal
(iii) the basic principle of signal detection and the circuitry designed to achieve the highest sensitivity.

The chapter begins with a brief description of the technology used to produce most biomedical devices.

2.1 Micro-Electro-Mechanical Systems

One of the major advances in biomedical devices technology over the past two decades has been the miniaturization of the sensor and its associated electrical circuitry (along with other components such as digital signal processing (DSP) chips). Lumped element electronic components such as resistors, inductors and capacitors are now usually produced using micro-electro-mechanical systems (MEMS) technology. Circuit diagrams throughout this book contain conventional symbols for resistors, capacitors, inductors and amplifiers, but whereas these used to be discrete components on the centimetre-size scale they are now often micro-machined and integrated in very small packages.

MEMS-based sensors have the advantages of small size, high sensitivity, high accuracy and high precision, as well as low power consumption. In biomedical

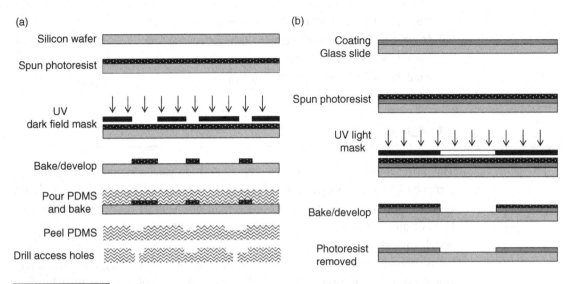

Figure 2.1   Schematics of simple processes to form MEMS devices. (a) Formation of a PDMS structure, often used in microfluidics. (b) A glass slide with specialized coating is used. Photoresist is spun onto the coating to form a very thin layer. Using a patterned mask, the UV light forms an imprint on the photoresist. The developing process removes the area of photoresist and coating that corresponds to the mask. Finally, the photoresist is removed by a chemical process to leave the coating with the geometry of the mask on top of the glass slide, which acts as a physical support.

devices they can be used to measure temperature, fluid/blood flow, pressure and acceleration, as well as many other parameters. In general, the small size also allows short response times with respect to rapid changes in the particular parameter being measured. Devices can be made of biocompatible substrates such as polydimethylsiloxane (PDMS), in addition to silicon, polyethylene and other inert compounds. The dimensions of MEMS-based biosensors range from micrometres up to several millimetres.

There are several textbooks and recent review articles [1–3] that give detailed accounts of MEMS fabrication methods, and so only a very brief description is given here. Many of the basic processes used to produce MEMS devices are similar to techniques used in the integrated circuit industry. Examples of these processes are shown in Figure 2.1 and include the following.

(i) **Photolithography** is a method in which a pattern generated by, for example, computer-aided design (CAD) software is transferred onto a glass mask. A substrate, spin-coated with photoresist (a photoresistive organic polymer), is placed in contact with the mask and both are illuminated with ultraviolet (UV) light, which makes the photoresist soluble. Finally, mask and substrate are separated, and the photoresist is removed.

(ii) Stencil lithography is a method that is based on vapour deposition technology. A pattern is produced using a shadow mask, corresponding to areas to be 'protected' from vapour deposition. The main advantages of this method are the submicrometre resolution and its applicability to fragile substrates such as biological macromolecules.

(iii) Thin films can be grown or deposited and can be made of silicon, plastics, metals and even biomolecules. Techniques such as sputtering and chemical vapour deposition can be used to produce these types of thin films.

(iv) Chemical etching can be used to selectively remove materials in desired patterns from a uniform starting pattern.

If the sensor is constructed of multiple parts, these need to be bonded together to form the final system. There are several methods of doing this including fusion bonding, which uses a chemical reaction between the molecules placed on the two surfaces; silicon-based systems can be thermally bonded at temperatures above 700°C, and silicon wafers can be bonded to glass substrates at much lower temperatures.

The first MEMS devices used in the biomedical industry were reuseable blood pressure sensors in the 1980s. Nowadays, MEMS-based pressure sensors represent the largest class of devices, including those for measuring blood pressure, intraocular pressure (IOP), intracranial pressure and intrauterine pressure. For example, disposable pressure sensors are used primarily during surgical procedures and in the intensive care unit (ICU) to monitor patient heart rate and blood pressure. During neonatal care this type of sensor is used to monitor the pressure around a baby's head during delivery. MEMS-based pressure sensors are also found in hospital equipment such as ventilators, drug infusion pumps and medication inhalers.

Devices for biomedical measurements are increasingly being formed on flexible substrates. One example is the concept of an 'epidermal electronic system' [4], in which devices can be conformally laminated onto the human skin to measure, for example, an ECG or electromyogram (EMG). Different approaches of forming these flexible devices are possible: one is to form elastomers such as PDMS, which is quite flexible due to its siloxane linkages. Conductive liquids, such as the gallium/indium/tin-based alloy Galinstan, can then be injected into elastic micro-channels formed in the PDMS. Alternatively, conductors such as gold can be patterned via thin-film deposition techniques, with prestretching of the flexible substrate performed to mitigate mechanical reliability issues. Liquid printing using modified ink-jet printers is also finding use in forming conducting elastomers. For example flexible antennas can be used for wireless transfer of data from implanted devices, as discussed in more detail in Chapter 8 [5].

An example of a MEMS device is shown in Figure 2.2, comprising a disposable contact lens with a strain-gauge element, an embedded loop antenna and an application-specific integrated circuit (ASIC) microprocessor (on a 2 × 2 mm chip). This sensor is used for monitoring of changes in corneal curvature that occur in response to changes in IOP in patients with glaucoma [6, 7], the second leading cause of blindness in the world after cataracts. A normal eye maintains a positive IOP in the range of 10 to 22 mmHg. Abnormal elevation (>22 mmHg) and fluctuation of IOP are considered the main risk factors for glaucoma, which can cause irreversible and incurable damage to the optic nerve. This initially affects the peripheral vision and can lead to blindness if not properly treated.

2.1.1 Noise in MEMS Devices

In any biomedical device there are two basic components of instrument noise: one is external noise (for example from power lines), which is coupled into the measurement system, and the second is the intrinsic noise of each component in the measurement device. Ultimately the overall noise level determines the smallest quantity that can be recorded. In MEMS and other devices [8–10], electronic noise can be separated into several components including:

(i) thermal/Johnson/Nyquist noise – this arises from temperature-induced fluctuations in carrier densities. The noise voltage over a measurement bandwidth BW is given by:

$$V_{\text{thermal noise}} = \sqrt{4kTR(BW)} \tag{2.1}$$

where k is Boltzmann's constant, T temperature in Kelvins, R resistance and BW in hertz.

(ii) shot noise, caused by the random arrival time of electrons (or photons), which can be modelled as a noise current:

$$I_{\text{shot noise}} = \sqrt{2qI_{dc}BW} \tag{2.2}$$

where I_{dc} is the average value of direct current in Amps, and q is the electron charge (1.6×10^{-19} C).

(iii) flicker (1/f) noise, which occurs due to variable trapping and release of carriers in a conductor, and can also be modelled as a noise current:

$$I_{\text{flicker noise}} = \sqrt{\frac{KI_{dc}BW}{f}} \tag{2.3}$$

where K is a constant that depends upon the type and geometry of the material used in the device.

(iv) phase noise in high-frequency RF systems (such as those used in wireless implanted devices), which arises from instabilities in oscillator circuits.

Mechanically based noise can also play a major role in MEMS devices, with one important example being Brownian motion effects on the proof-mass in an accelerometer. As shown by Gabrielson [10], the spectral density of the fluctuating force related to a mechanical resistance is given by a direct analogy of the Johnson noise for electrical resistance:

$$F_{\text{mechanical-thermal noise}} = \sqrt{4kTR_{mech}} \tag{2.4}$$

where R_{mech} is the mechanical resistance, or damping coefficient, of the device. In order to reduce the noise, the sensor structure should be designed with as large a proof-mass and small a damping factor as possible [11].

Which type of noise dominates depends very much on the structure of the MEMS device. In general, the thermal or mechanical-thermal terms are the largest ones, with their contributions increasing as devices become smaller with thinner metallic conductors.

2.2 Voltage Sensors: Example – Biopotential Electrodes

The most commonly used voltage sensors in the clinic are electrodes placed on the skin to measure potential differences between different regions of the body. These measurements reflect cellular or neural electrical activity in organs within the body. The classic example is the ECG. Other measurements include the electro-encephalogram (EEG), electrooculogram (EOG) and electromyogram (EMG).

2.2.1 Clinical and Biomedical Voltage Measurements

As stated above, the ECG is one of the most common clinical measurements and involves external monitoring of potential differences between surface electrodes placed on the body. Heart disease represents one of the most common causes of hospitalization and death, particularly in the Western world. The heart is essentially a large muscle that pumps blood around the body. Muscle contractions caused by action potentials (see next section) within the cardiac muscle cells perform the pumping action. Any deficiencies in the propagation of action potentials through the heart can cause a serious health risk. This is true not only in everyday life, but also during surgery and other medical procedures. Action potentials cause ionic currents to flow throughout the body, and these currents cause a voltage to develop at the body surface. These voltages can be measured as potential differences between electrodes placed at different positions on the skin surface. The ECG, covered in more detail in Chapter 5, is a measurement that detects voltage waveforms highly characteristic of different types of heart diseases. Other types of voltage measurements include EEG recordings, which are performed using similar principles to ECG, and are often used during surgery to monitor the depth of anaesthesia.

2.2.2 Action Potentials and Cellular Depolarization

Electrical signals measured by electrodes on the skin surface arise from action potentials that occur at the cellular level, and set up ionic current paths within the body. Considering first cells in muscle and brain tissue (cardiac action potentials are described later in this chapter), the intracellular and extracellular concentrations of the major cations and anions when the cell is at rest are approximately:

Intracellular Na^+: 12 mM Extracellular Na^+: 145 mM
Intracellular K^+: 155 mM Extracellular K^+: 4 mM
Intracellular Cl^-: 4 mM Extracellular Cl^-: 120 mM

There is clearly a large imbalance between the intracellular and extracellular ionic concentrations. In order to maintain this steady-state concentration gradient across the membrane, some form of active transport is required. This transport is provided by the **sodium–potassium pump**, which, powered by hydrolysis of adenosine triphosphate (ATP), actively transports Na^+ from the intracellular to the extracellular space, and K^+ in the opposite direction.

The difference in intracellular and extracellular concentrations results in a resting potential, E, across the membrane, which is given by the Goldman–Hodgkin–Katz (GHK) equation:

$$E = \frac{RT}{F} \ln \left\{ \frac{P_k[K^+]_o + P_{Na}[Na^+]_o + P_{Cl}[Cl^-]_i}{P_k[K^+]_i + P_{Na}[Na^+]_i + P_{Cl}[Cl^-]_o} \right\} \qquad (2.5)$$

where P is the permeability of the membrane to the particular ionic species, the subscripts o and i refer to extracellular and intracellular, respectively, R is the gas constant (8.31 J/(mol.K)), F is Faraday's constant (96 500 C/mole) and T the absolute temperature in kelvins.

Skeletal muscle and ventricular and atrial cardiac muscle typically have resting potentials in the region of –85 to –90 mV, whereas nerve cells have resting potentials of about –70 mV. In the former case, the cell membrane permeability of sodium ions is $\sim 2 \times 10^{-8}$ cm/s, that of potassium ions $\sim 2 \times 10^{-6}$ cm/s and that of chloride ions 4×10^{-6} cm/s. Using these values, the GHK equation results in the resting membrane potential being approximately –85 mV.

A cell with a negative resting potential is said to be **polarized**. Figure 2.3 shows the changes that occur in a nerve cell if an electrical stimulus is provided to the cell. The first thing that happens is that the (trans)membrane potential becomes more positive. The more positive membrane potential causes an increase in the permeability of the cell membrane to Na^+. The intracellular Na^+ concentration increases, the extracellular Na^+ concentration decreases and the transmembrane potential becomes even more positive. If the membrane potential exceeds a threshold level, approximately –65 mV, then an **action potential** occurs, as shown in Figure 2.3. A certain time after the membrane potential begins to rise, a counteracting mechanism starts in which the membrane permeability to K^+ also increases. This results in a flow of K^+ from the intracellular to the extracellular medium, acting to make the membrane potential more negative. The Na^+ permeability returns to its resting value within a millisecond, and that of potassium slightly slower, such that after a few milliseconds the membrane potential has returned to its equilibrium value. In fact, a hyperpolarizing after-potential can be seen in Figure 2.3, in which the membrane potential is more negative than the resting potential. This results from the relatively slow closing of the K^+ channels.

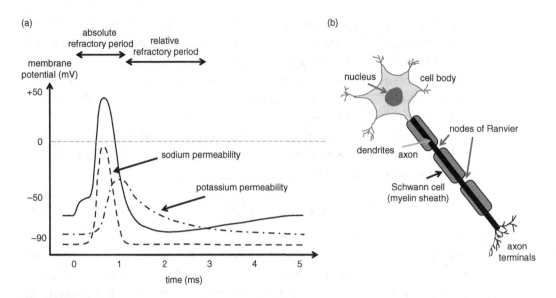

Figure 2.3   (a) Plot of the cell membrane potential in a nerve cell as a function of time, corresponding to an action potential. A rise in sodium permeability slightly precedes that of potassium. The absolute refractory period lasts for approximately 1 ms, with the relative refractory period being an additional 2 ms. (b) The action potential at one node of Ranvier depolarizes the membrane at the next node, thus propagating the action potential down the axon.

The action potential is an **all-or-nothing event**; if the stimulus is sub-threshold then no depolarization occurs, and if it is above the threshold then the membrane potential will reach a maximum value of around +50 mV irrespective of the strength of the stimulus. There are also two different types of refractory period, absolute and relative, which are shown in Figure 2.3. During the absolute refractory period, if the cell receives another stimulus, no matter how large, this stimulus does not lead to a second action potential. During the relative refractory period, a higher than normal stimulus is needed in order to produce an action potential.

As indicated earlier, there are slight differences in the action potentials of different types of cells. For example, skeletal muscle cells have a depolarizing after-potential, which is different than the hyperpolarizing after-potential that is characteristic of most neuronal action potentials, as seen in Figure 2.3. As covered above, for nerve and skeletal muscle cells, it is the opening of voltage-gated Na^+ channels (increasing the Na^+ permeability) that produces rapid depolarization. This is also true for ventricular and atrial cardiac muscle cells. However, for the sinoatrial and atrioventricular nodes in the heart, and most smooth muscle cells, it is the opening of voltage-gated Ca^{2+} channels that produces a more gradual depolarization.

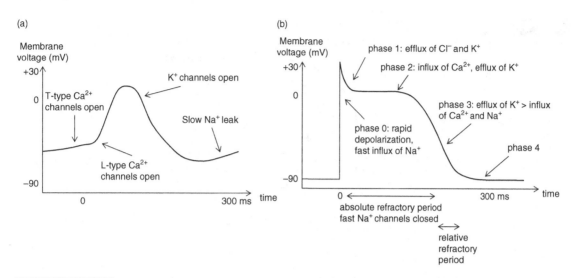

Figure 2.4 Depolarization and repolarization processes in (a) cardiac pacemaker cells and (b) cardiac ventricular cells. L-long lasting, T-transient

Action potentials do not occur at single isolated locations, but instead propagate from multiple positions, which allows their detection at the body surface as described in the next section. The speed of propagation is between 1 m/s and 100 m/s. Propagation of action potentials occurs via different mechanisms depending upon which type of cells are involved. In nerve cells the conduction of action potentials is termed salutatory conduction, and basically consists of excitation of successive nodes of Ranvier, which lie in between myelinated segments of the axon, as shown in Figure 2.3(b). For a skeletal muscle fibre, an action potential propagates away from the neuromuscular junction in both directions toward the ends of the fibre.

In the heart, the action potential is initiated by specialized cells in the sinoatrial (SA) node called pacemaker cells. In these cells, as already mentioned, Ca^{2+} plays an important role. At rest, it is present within the myocardial cell at an intracellular concentration of 10 mM and at a concentration of 2 mM in the extracellular fluid. The corresponding depolarization process is shown in Figure 2.4(a). The action potential rapidly spreads throughout the rest of the heart, although the action potential in the atria (as well as in the atrioventricular (AV) node, Purkinje fibres and ventricles) is of different size, shape and duration than that of the SA node. Figure 2.4(b) shows the form of the action potential in ventricular cells. Notice that the timescales associated with repolarization of cardiac cells are quite different from those for neuronal activation. There is an extensive plateau period in which the membrane potential is at a high positive voltage, which arises from the relatively slow timescales associated with calcium channels opening.

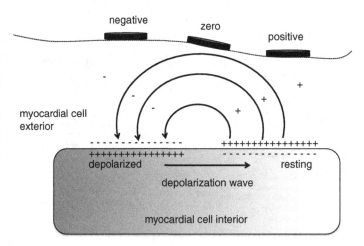

Figure 2.5

Illustration of the propagation of ionic currents produced by a depolarization wave spreading via the heart, through a homogeneous volume conductor and the corresponding surface biopotentials measured by surface electrodes.

2.2.3 Surface Electrode Design

The propagation of action potentials along axons or muscle fibres produces ionic currents that flow through the body, as shown in Figure 2.5 for the case of myocardial cells in the heart. Electrodes placed on the skin surface are used to convert the ionic currents within the body into an external potential (or potential difference between electrodes). These types of electrodes are used for measurements such as ECG, EEG, EMG and EOG. The relationship between the potentials that exist at the body surface and the underlying sources of electrical activity can be determined using the concept of volume conductor fields. The term volume conductor refers to the body part through which the ionic currents propagate. Many different types of volume conductor can be used to model the propagation effects, ranging from very simple models assuming a single homogeneous tissue type, to much more realistic models with accurate geometries of different organs and tissues, with appropriate physical parameters [12, 13]. Using these models, the propagation paths of the ionic currents and the potentials created at different points on the body surface can be estimated.

The electrodes used to measure these biopotentials are formed from a metal in contact with an electrolyte: in electrochemistry this arrangement is referred to as a half-cell. When the metal first comes into contact with the electrolyte a metal atom in the electrode dissolves and is transported as a positive ion across the naturally occurring Helmholtz double layer (consisting of a layer of ions tightly bound to the surface of the electrode and an adjacent layer of oppositely charged ions in solution), causing the electrolyte to acquire a net positive charge while the electrode acquires a net negative charge. As the potential at the electrode increases over time, due to this transport process, it creates a correspondingly stronger electric field within the double layer. The potential continues to rise until this electric field stops the transport process. At this point the potential is referred to as the half-cell potential (V_{hc}). The value of V_{hc} depends upon the composition of the electrode and the surrounding electrolyte.

The most commonly used electrode for biomedical measurements and devices is silver/silver chloride (Ag/AgCl) because of its low half-cell potential of approximately 220 mV and its ease of manufacture. In the form used for ECG and EEG, a silver disc is covered by a thin layer of silver chloride deposited on its surface. The electrode is placed in a gel that contains a saturated solution of sodium chloride. When the electrode comes into contact with a solution that contains chloride ions, an exchange takes place between the electrode and the solution. The reactions at the electrode and within the electrolyte are given, respectively, by:

$$Ag(\text{solid}) \leftrightarrow Ag^+(\text{solution}) + e^-(\text{metal})$$

$$Ag^+(\text{solution}) + Cl^-(\text{solution}) \leftrightarrow AgCl(\text{solid})$$

which gives an overall reaction:

$$Ag(\text{solid}) + Cl^-(\text{solution}) \leftrightarrow AgCl(\text{solid}) + e^-(\text{metal})$$

The silver–silver chloride reference electrode develops a potential E, which depends upon the chloride concentration in the electrolyte:

$$E = E^\varnothing - \frac{RT}{F} \ln a_{Cl^-} = 0.22 - 0.059\log[Cl^-] \tag{2.6}$$

where $E^\varnothing$ is the standard half-cell potential, R is the gas constant, T the temperature in kelvins, F Faraday's constant and a_{Cl^-} represents the activity of the chloride ion.

Figure 2.6 shows the passage of ionic currents, created by (neuronal or cardiac) electrical activity, through the skin and into the electrolyte. In this way, the ionic current is transduced into an electric signal by the electrode. The electrical circuit equivalents of different tissues, materials that make up the electrode and interfaces between them, are also shown. The skin is represented as having three different layers. The dermis and subcutaneous layer are modelled as a resistor. The next layer towards the surface is the epidermis, which is semi-permeable to ionic current, and can be modelled as a parallel RC circuit. The very top of the three layers making up the epidermis is called the stratum corneum, and consists of dead material that is typically scrubbed with abrasive tissue to remove the layer before placing the electrode on the skin surface. In order to accurately measure the potential difference between the two electrodes, the contact impedance between the skin and the electrode (R_{s-e} in Figure 2.6) should be between 1 kΩ and 10 kΩ. The electrolyte

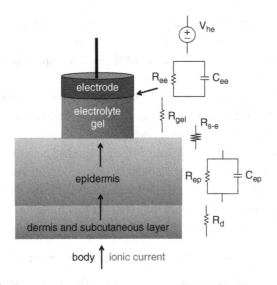

Figure 2.6 An electrical circuit model of the body/skin/electrolyte/electrode system corresponding to the transduction of ionic currents in the body to an electrical signal from the electrode. V_{he} is the half-cell potential, R_{ee} and C_{ee} represent the electrode/electrolyte gel interface, R_{gel} the resistance of the electrolyte gel, R_{s-e} the gel/skin interface, R_{ep} and C_{ep} form a parallel circuit to model the epidermis, and the dermis and subcutaneous layer have resistance R_d.

can be modelled as a simple resistor. The double layer at the electrode–electrolyte interface corresponds to a parallel RC circuit, in series with the half-cell potential, V_{he}. The electrolytic gel serves several purposes. Most importantly, it enables the transfer of charge from the tissue to the electrode. In addition, it increases the actual contact area between the skin and electrodes, since the skin does not have a perfectly flat surface, which results in reduced contact impedance. It also diffuses into the skin, which further reduces the contact impedance, and allows minor movements of the electrode without a loss of contact.

For ECG recording, the most common Ag/AgCl electrode is a disposable one that is 'pre-gelled'. A silver-plated disc is connected to a snap-connector, as shown in Figure 2.7, which is then connected to the ECG lead. The disc is plated with AgCl, with an electrolyte gel containing sodium chloride covering the disc. There is an adhesive ring around the electrode, which can be pressed onto the patient's skin.

The amplitude of an ECG biopotential is typically a few mV, with components covering a frequency range from DC to roughly 30 Hz. However, there is a very large offset voltage which occurs from 50/60 Hz interference from electricity sources and power lines coupling to the ECG leads and the body itself, and also from the DC half-cell potential at the electrodes. For EEG measurements the biopotentials have an amplitude in the tens of microvolts over a frequency range of DC to 50 Hz, with

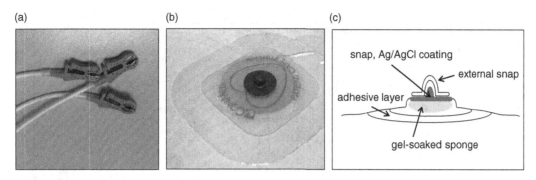

(a)   (b)   (c)

Figure 2.7   (a, b) Photographs of ECG leads, which are connected to disposable stick-on electrodes. (c) Schematic of a disposable Ag/AgCl electrode.

interference from ECG, EMG and EOG signals, as well as 50/60 Hz pick-up. The design of amplifiers and filters for subsequent steps in acquisition of ECG and EEG signals is considered in Chapters 5 and 6, respectively.

2.3 Optical Sensors: Example – a Pulse Oximeter

Fibre-optic sensors can be used for a number of different types of measurements, including pressure and displacement. Interferometric principles (Problem 2.4) or changes in light intensity produced by displacement of a flexible diaphragm can both be used. However, this section concentrates on the most commonly encountered medical optical sensor, the pulse oximeter, which is a device used to measure blood oxygenation non-invasively, and can also be used to monitor the patient's heart rate. Pulse oximeters are very small, inexpensive devices that can be placed around a finger, toe or ear lobe, and can be battery operated for home monitoring. The **optical sensor** consists of two photodiodes, one emitting light in the red part of the electromagnetic spectrum and the other in the infrared part of the spectrum. Light that passes through the tissue to a photodiode detector placed on the other side is converted into a voltage, which is then filtered, amplified and digitized.

2.3.1 Clinical Blood Oxygenation Measurements

It is critical to be able to monitor the level of blood oxygenation in patients during surgery, anaesthesia and other invasive interventions. Oxygen is required for oxidative phosphorylation and aerobic cellular metabolism, since it functions as an electron acceptor in mitochondria to generate chemical energy. Oxygen is

(a)

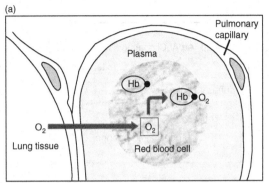

(b)

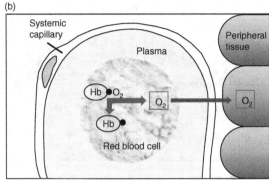

Figure 2.8 Schematics of (a) the incorporation of oxygen into the haemoglobin complex, which occurs in the lungs; and (b) the delivery of oxygen from the red blood cell to tissue in different parts of the body.

inhaled into the lungs and diffuses into red blood cells (RBCs) in the capillaries via the alveoli. Once inside the RBCs the oxygen becomes chemically, but reversibly, bound to one of the four haem groups in each haemoglobin molecule: the complex is referred to as oxyhaemoglobin (HbO_2). Oxygenated blood flows to the heart, where it is pumped via the systemic arterial system to various organs of the body (Figure 2.8). When the oxyhaemoglobin reaches the cells within a particular organ, it is reduced to its deoxygenated form, deoxyhaemoglobin (Hb), releasing oxygen and also producing carbon dioxide (CO_2) and water. The CO_2 becomes bound to the Hb molecule to form carbaminohaemoglobin. Systemic blood flow returns the RBCs to the lungs where the CO_2 is exhaled. The average circulation time of an RBC in the body is approximately 30 to 45 seconds.

Normal blood oxygenation levels are quantified in terms of the saturation of peripheral oxygen, SpO_2. The value of SpO_2 represents the percentage of haemoglobin in the arteries that is oxygenated, and is given by:

$$SpO_2 = \frac{[HbO_2]}{[HbO_2] + [Hb]} \tag{2.7}$$

where $[HbO_2]$ is the concentration of oxyhaemoglobin, and $[Hb]$ the concentration of deoxyhaemoglobin. Typical values of SpO_2 in healthy people are between 95 and 100%, corresponding to a full oxygen load of the haemoglobin molecules. If the SpO_2 falls below 90% the condition is referred to as hypoxaemia, and if the level is below 80% for a prolonged period then there is serious danger of organ failure. Low SpO_2 can be caused by problems in the lungs resulting in insufficient perfusion of oxygen to the blood, or in damage to the blood cells themselves as in the case of anaemia. In newborn babies, low SpO_2 can be indicative of conditions

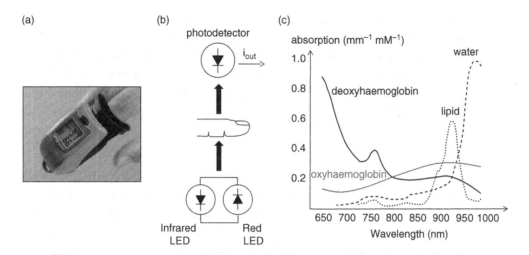

(a)

(b)

(c)

Figure 2.9 (a) Photograph of a pulse oximeter placed around a patient's finger. (b) The principle of measurement of blood oxygenation using a pulse oximeter involves the transmission of red and infrared light through the finger to a photodetector. The output current from the photodetector is proportional to the intensity of transmitted light. (c) Wavelength-dependent absorption of oxyhaemoglobin, deoxyhaemoglobin, water and lipid. Red light corresponds to a wavelength of approximately 660 nm and infrared light to 940 nm.

such as septicaemia. Other diseases that cause low SpO_2 in adults are congenital heart defects, anaemia and respiratory distress syndrome.

2.3.2 Measurement Principle Using an Optical Sensor

A pulse oximeter, shown in Figure 2.9, consists of a probe that clips on to a patient's ear lobe, toe or finger. Light-emitting diodes (LEDs) on one side of the device shine red and infrared light through the ear lobe, toe or finger to a photodiode detector located on the other side. The pulse oximeter estimates blood oxygenation by measuring the amount of light transmitted through the tissue, and using the different infrared and red-light absorption properties of Hb and HbO_2, which are shown as a function of wavelength in Figure 2.9. Also shown for reference are the frequency-dependent absorption coefficients of water and lipid in tissue.

In order to calculate the SpO_2 in equation (2.7), the concentrations of both HbO_2 and Hb must be determined. These concentrations, C, can be derived from the Beer–Lambert law:

$$I = I_0 e^{-\alpha C d} \tag{2.8}$$

where I is the measured light intensity, I_0 is the incident light intensity, α is the absorption coefficient and d is the path length of the light through the tissue. The intensity of the incident light I_0 is assumed to be proportional to the current through the LED, and the current generated by the photodiode to be proportional to the intensity of the transmitted light. The measured intensity, I_R, of transmitted red light at 660 nm is given by:

$$I_R = I_{0,R}e^{-\alpha_{R,tissue}[tissue]d - \alpha_{R,Hb}[Hb]d - \alpha_{R,HbO_2}[HbO_2]d} \qquad (2.9)$$

One complication is that on a heartbeat by heartbeat basis, a small amount of arterial blood is pumped into the tissue, and this blood then slowly drains back through the venous system. Therefore the amount of light that passes through the tissue varies as a function of the time-dependent blood volume. This factor can be accounted for by taking the ratio of the maximum to minimum detected intensities over the duration of a heartbeat to eliminate the absorption effects of the surrounding stationary tissues.

$$\frac{I_{R,max}}{I_{R,min}} = e^{\alpha_{R,HbO_2}[HbO_2](d_2-d_1) + \alpha_{R,Hb}[Hb](d_2-d_1)} \qquad (2.10)$$

where d_1 and d_2 refer to the very slightly different tissue dimensions through which the light travels over the heart cycle. An equivalent equation can be written for the transmitted near infrared signal, I_{IR}:

$$\frac{I_{IR,max}}{I_{IR,min}} = e^{\alpha_{IR,HbO_2}[HbO_2](d_2-d_1) + \alpha_{IR,Hb}[Hb](d_2-d_1)} \qquad (2.11)$$

The red-to-infrared pulse modulation ratio, R, is defined as:

$$R = \frac{\ln\left(\frac{I_{R,max}}{I_{R,min}}\right)}{\ln\left(\frac{I_{IR,max}}{I_{IR,min}}\right)} = \frac{\alpha_{R,HbO_2}[HbO_2] + \alpha_{R,Hb}[Hb]}{\alpha_{IR,HbO_2}[HbO_2] + \alpha_{IR,Hb}[Hb]} \qquad (2.12)$$

Equations (2.7) and (2.12) can now be combined to give:

$$SpO_2 = \frac{\alpha_{R,Hb} - \alpha_{R,HbO_2}R}{\alpha_{R,Hb} - \alpha_{R,HbO_2} + \left(\alpha_{R,HbO_2} + \alpha_{IR,Hb}\right)R} \qquad (2.13)$$

In most commercial pulse oximeters, SpO_2 is assumed to follow a quadratic polynomial in R.

$$SpO_2 = a + bR + cR^2 \qquad (2.14)$$

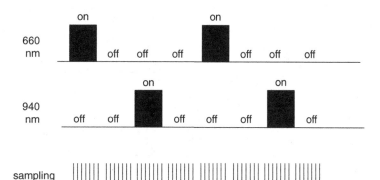

Figure 2.10

Timing diagram showing the two LEDs alternately switched on for 250 μs, with a 250 μs off-time to measure the stray light component, which is subtracted in subsequent signal processing.

where the coefficients a, b and c are obtained by fitting a second-order polynomial to a graph of measured R values vs. SpO_2 levels from a previously calibrated pulse oximeter for different test subjects. Calibration of individual pulse oximeters is necessary since there may be significant batch-to-batch variations in the emitter wavelength characteristics of the LEDs.

2.3.3   Optical Transmitter and Detector Design

The two LEDs, with peak outputs at wavelengths of 660 nm (red) or 940 nm (infrared), are typically high power gallium-aluminium-arsenide (GaAlAs) emitters. The power output is slightly lower for the infrared LED (2 mW) vs. 2.4 mW for the red LED for an input current of 20 mA. The LEDs are rapidly and sequentially excited by two current sources, one for each LED: typical driving currents are 10 mA for the red LED and 15 mA for the infrared LED. Turn-on times are on the order of 1 microsecond or less, and the LED is turned off by applying a reverse voltage. Depending upon the packaging the LEDs can be back-to-back in parallel, or in series with a common ground and separate driving inputs. The detector is synchronized to capture the light from each LED as it is transmitted through the tissue, as shown in Figure 2.10.

The detector is typically a photodiode that uses the photovoltaic effect, i.e. the generation of a voltage across a positive–negative (P–N) junction of a semiconductor when the junction is exposed to light. A silicon planar-diffused photodiode has output characteristics shown in Figure 2.11. Since the responsivity is higher at 940 nm than at 660 nm, the current driving the red LED is higher than for the infrared LED, as described previously.

The P–N junction in a photodiode operates as a transducer converting light into an electrical signal. The spectral response can be adjusted by varying the thickness

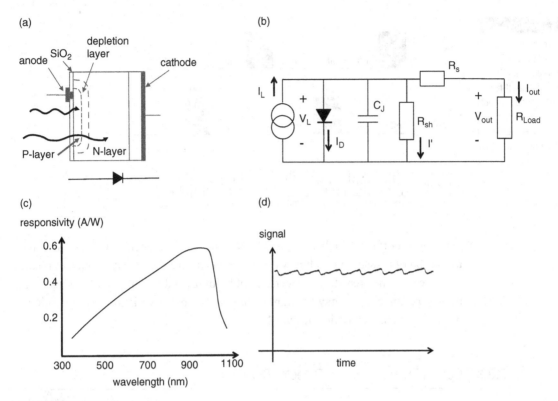

Figure 2.11 (a) Schematic of a semiconductor photodiode detector. (b) Circuit representation of the photodiode. I_L is the current produced by the incident light, I_D is the diode current, C_j is the junction capacitance, R_{sh} the shunt resistance, R_s the series resistance, I' the shunt resistance current, and I_{out} the output current that passes to the filters and amplifiers in the next stage of the device. (c) Plot of the responsivity, i.e. the output current in amps per input power in watts, as a function of wavelength for a typical photodiode. (d) Schematic of the raw signal from one channel of the pulse oximeter, illustrating a small AC signal on top of a much larger DC offset.

of the outer P-layer and inner N-layer. If the energy of the light is greater than the bandgap energy E_g, electrons are excited into the conduction band, leaving holes in the valence band. This means that the upper wavelength is set by:

$$\lambda < \frac{1240}{E_g} nm \tag{2.15}$$

For silicon devices the bandgap energy is 1.12 eV, corresponding to an upper wavelength of 1100 nm, as indicated in Figure 2.11. For lower wavelengths (corresponding to high frequency and high energy) then light absorption within the diffusion layer is very high. The number of electron-hole pairs is proportional to the amount of incident light, resulting in a positive charge in the P-layer and a negative one in the N-layer. The thinner the diffusion layer and the closer the P–N junction is to the surface, the higher is the sensitivity of the photodiode. The detection limit is determined by the total noise of the detector, which is the sum of the thermal noise from the shunt resistance (R_{sh}) of the device and the shot noise arising from the dark current and photocurrent. The effective cut-off frequency is between 300 and 400 nm for normal photodiodes.

Using the circuit representation in Figure 2.11(b) the output current is given by:

$$I_{out} = I_L - I_D - I' \quad = I_L - I_S\left(e^{\frac{eV_D}{kT}} - 1\right) - I' \tag{2.16}$$

where V_D is the applied bias voltage, I_S is the photodiode reverse saturation current, e is the electron charge, k is the Boltzmann constant and T is the absolute temperature of the photodiode. As will be covered in section 3.6.1 the output of the photodiode is connected to a transimpedance amplifier, which has a very low input impedance. Therefore the appropriate output current can be calculated by setting the load resistance to zero, in which case the short-circuit current can be derived from:

$$I_{sc} = I_L - I_S\left(e^{\frac{e(I_{sh}R_S)}{kT}} - 1\right) - \frac{I_{sh}R_s}{R_{sh}} \tag{2.17}$$

The last two terms on the right-hand side represent non-linear terms, but considering that R_{sh} is typically 10^7 to 10^{11} times larger than R_s, these terms effectively disappear, and the short-circuit current is linearly dependent on the light intensity over many orders of magnitude.

As can be seen from Figure 2.11(d) the current from the photodiode has a very large DC component and a much smaller AC component (~1% of the DC signal). As outlined earlier, within the path of the light there is a lot of tissue that does not contain blood and scattered light from this tissue gives rise to the DC signal. The AC signal contains the desired signal coming from the blood, as well as components from noise sources such as 50/60 Hz light pollution. Electronic circuits to remove these interfering signals, as well as signal amplification are considered in section 3.6.1.

2.4   Displacement/Pressure Sensors and Accelerometers

The measurement of pressure, as well as changes in pressure and tissue displacement associated with conditions such as tissue swelling, has many very important clinical applications. Many of the measurements require invasive insertion of a pressure sensor, and so small dimensions and a high degree of biocompatibility are key elements of the sensor design. There are many types of sensor that can be used

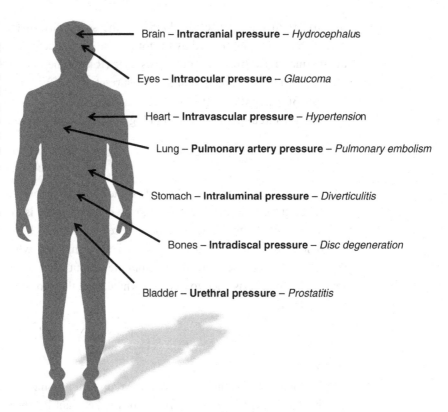

Figure 2.12

Examples of different organs in which pressure measurements are performed clinically, and corresponding pathological conditions that can cause changes in pressure compared to normal values.

Brain – **Intracranial pressure** – *Hydrocephalus*

Eyes – **Intraocular pressure** – *Glaucoma*

Heart – **Intravascular pressure** – *Hypertension*

Lung – **Pulmonary artery pressure** – *Pulmonary embolism*

Stomach – **Intraluminal pressure** – *Diverticulitis*

Bones – **Intradiscal pressure** – *Disc degeneration*

Bladder – **Urethral pressure** – *Prostatitis*

to measure pressure and displacement. These include: (i) (piezo)resistive, (ii) piezoelectric, (iii) capacitive and (iv) inductive sensors. Almost all of these are produced using MEMS technology, and contain integrated bridge measurement circuits as described in section 2.4.2.1.

2.4.1   Clinical Pathologies Producing Changes in Internal Pressure

Figure 2.12 shows a range of different clinical conditions that cause internal pressure changes to occur. Measuring and monitoring these pressures is an important component of diagnosis and assessing the progress of treatment.

2.4.2   Resistive and Piezoresistive Transducers

Measurements of physical displacement can be performed using a strain gauge, which is defined as a device whose resistance depends upon the strain (given by the fractional change in the dimensions of the device) imposed upon it. The principle is very simple: for example, when a thin wire is stretched its resistance changes. A wire with circular cross-sectional area A, diameter D, length L and resistivity ρ, has a resistance R given by:

$$R = \frac{\rho L}{A} \tag{2.18}$$

where $A = \pi D^2/4$. The strain is defined as $\Delta L/L$ (usually expressed in units of parts per million, ppm). A change in length ΔL produces a change in resistance ΔR given by:

$$\frac{\Delta R}{R} = \frac{\Delta \rho}{\rho} + \frac{\Delta L}{L} - \frac{\Delta A}{A} \tag{2.19}$$

As the length of the wire changes, so does its diameter. The Poisson's ratio, μ, of the material relates the change in length to the change in diameter:

$$\mu = -\frac{\Delta D}{D}\frac{L}{\Delta L} \tag{2.20}$$

The change in resistance from equation (2.19) can be separated into two terms:

$$\frac{\Delta R}{R} = (1 + 2\mu)\frac{\Delta L}{L} + \frac{\Delta \rho}{\rho} \tag{2.21}$$

The first term on the right-hand side is called the 'dimensional term' and the second the 'piezoresistive term'. Equation (2.21) can also be expressed in terms of a material's gauge factor, G, which is defined as the fractional change in resistance divided by the strain. The value of G can be used to compare the characteristics of different materials that might be used for the thin wires:

$$G = \frac{\Delta R/R}{\Delta L/L} = (1 + 2\mu) + \frac{\Delta \rho/\rho}{\Delta L/L} \tag{2.22}$$

Metals have gauge factors ranging from 2 to 5. Semiconductors such as silicon have values more than an order of magnitude higher, mainly due to the additional piezoresistive term, and these latter types of sensors are referred to as piezoresistive sensors or piezoresistors. The reason why semiconductors have a high piezoresistive term can be seen from the expression for the resistivity:

$$\rho = \frac{1}{nq\mu'} \tag{2.23}$$

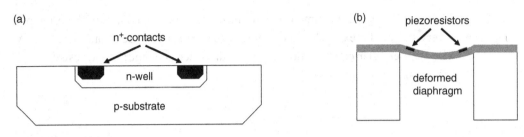

Figure 2.13   (a) Schematic cross-section of the basic elements of a silicon-based n-well piezoresistor. (b) The piezoresistors are typically placed at the points of maximum deflection of the diaphragm for the highest sensitivity.

where n is the number of charge carriers, q is the charge per single charge carrier and μ' is the mobility of charge carriers (the prime accent has been added to differentiate from Poisson's ratio above). The value of μ' is related to the average atomic spacing in the lattice, and therefore changes with applied pressure or physical deformation: as a result the resistivity also changes. Barlian *et al.* [14] have published a review of developments in semiconductor piezoresistors.

One example of a biomedical piezoresistive sensor is shown in Figure 2.13. The pressure-sensing element combines resistors and an etched diaphragm structure to produce a structure with a pressure-dependent resistance [15]. When incorporated into a bridge circuit, discussed in the next section, this change in resistance can be transduced into a change in the output voltage of the circuit. Miniature strain gauges produced using MEMS technology can also be implanted into the body to study, for example, movement close to bone implants and to monitor small cracks in the bone.

2.4.2.1   Bridge Circuits

Bridge circuits are used to convert a change in the impedance (whether resistance, capacitance or inductance) of a sensor into a change in an output voltage, which can subsequently be filtered, amplified and digitized. For example, the strain gauge described in the previous section is typically integrated into a **Wheatstone bridge resistor network** in order to produce a voltage output that is proportional to the change in resistance. The Wheatstone bridge circuit, shown in Figure 2.14, consists of four resistors, three with fixed values and the fourth which is the (piezo) resistive sensor.

Applying voltage division to the circuit shown in Figure 2.14 gives the transfer function for V_{out}/V_{in}.

Figure 2.14

Circuit diagram of a resistive Wheatstone bridge. Resistors R_1, R_2 and R_3 are fixed (and usually have the same value) and R_4 represents the (piezo)resistive sensor whose resistance changes with pressure or displacement.

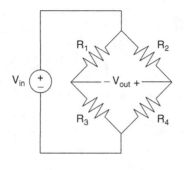

$$\frac{V_{out}}{V_{in}} = \frac{R_2 R_3 - R_1 R_4}{(R_1 + R_3)(R_2 + R_4)} \qquad (2.24)$$

In the case that $R_1 = R_2 = R_3 = R_4 = R$, and a change in pressure increases the value of R_4 to $R + \Delta R$, the change in the output voltage is given by:

$$\Delta V_{out} = -\frac{R \Delta R}{4R + \Delta R} V_{in} \qquad (2.25)$$

If $\Delta R \ll R$, then there is a linear relationship between the change in the output voltage and the change in resistance:

$$\Delta V_{out} = -\frac{\Delta R}{4} V_{in} \qquad (2.26)$$

The output becomes non-linear with respect to ΔR if the change is large. There are a number of variations on the basic Wheatstone bridge that can overcome this limitation, for example the Anderson bridge (see Problems). Although many of the original designs for biosensors used macroscopic resistor networks, Wheatstone and other bridge circuits are now typically integrated as microfabricated elements: as an example, a microfabricated Wheatstone bridge piezoresistive silicon pressure sensor, with dimensions of a few millimetres, is shown in Figure 2.15.

2.4.2.2 Applications of Resistive Sensors in Venous Occlusion Plethysmography and Disposable Blood Pressure Monitors

Venous occlusion plethysmography (VOP) is a simple technique that is used to assess the general vascular function of the patient. It can also be used to measure the effects of systemically administered drugs such as vasodilators. In VOP a cuff is inflated around the lower leg (the forearm can also be used) to a pressure of approximately 40 mm Hg so that arterial inflow into the lower leg is not affected, but venous flow is stopped, as shown in Figure 2.16. The volume of the leg therefore increases, with the rate of volume increase equal to the blood flow. A fine elastic tubular hose filled with a gallium–indium alloy, sealed at both ends, with platinum contacts is placed around the lower leg. The increase in volume is reflected by a change in the resistance of the strain gauge as described in equation (2.21). The strain gauge is integrated into a Wheatstone bridge circuit, and the change in calf volume can be directly related to the blood flow. In a similar technique, piezoresistive strain gauges are also used in cardiovascular and respiratory measurements of the volume and

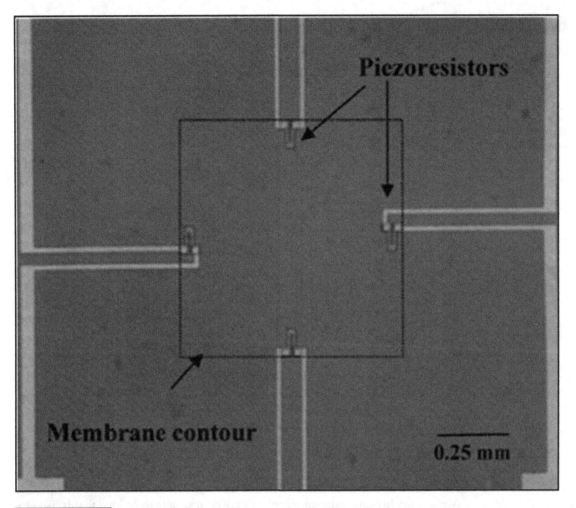

Figure 2.15 Four microfabricated piezoresistors placed at maximum stress locations on a flexible substrate, and integrated into a bridge measurement circuit. Figure reproduced from Wu *et al.* [16].

dimensions of tissues, such as vessel diameter, by placing a strain gauge around the particular vessel.

Piezoresistive sensors are also extensively used in disposable blood-pressure sensors, such as the one shown in Figure 2.17.

2.4.3 Piezoelectric Sensors

Piezoelectric materials have the property that they generate a voltage when a mechanical force is applied to one face of the material, with the voltage being

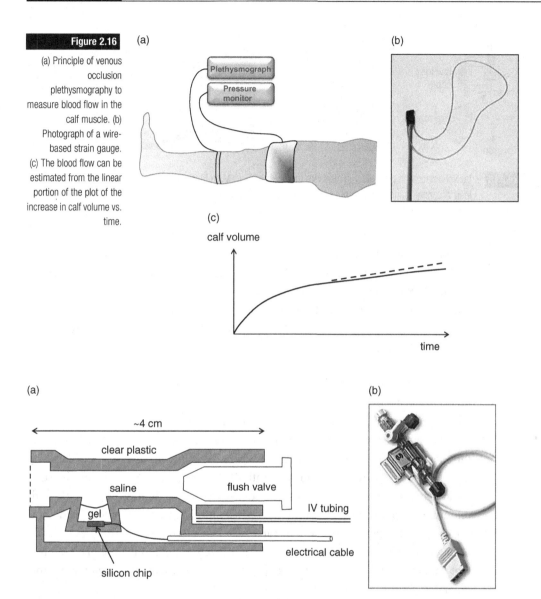

Figure 2.16

(a) Principle of venous occlusion plethysmography to measure blood flow in the calf muscle. (b) Photograph of a wire-based strain gauge. (c) The blood flow can be estimated from the linear portion of the plot of the increase in calf volume vs. time.

Figure 2.17   (a) Schematic drawing of a disposable blood-pressure sensor placed in-line with an intravenous saline bag, which prevents clotting. The silicon chip has an integrated Wheatstone bridge configuration, with a protective gel covering all electrical connections in case cardiac defibrillation is applied. (b) Photograph of a disposable blood-pressure sensor.

directly proportional to the applied force. Examples of piezoelectric materials include polyvinylidene difluoride (PVDF) and quartz. Unlike piezoresistive strain-gauge sensors, **piezoelectric sensors** require no external voltage to produce a signal. Piezoelectric sensors typically have a very high output impedance

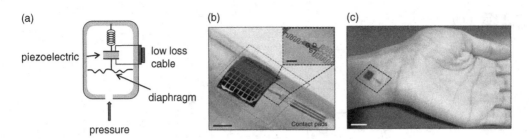

Figure 2.18 (a) Schematic of a piezoelectric pressure transducer. (b, c) Microfabricated piezoelectric sensors made from lead zirconate titanate, which are flexible and can be attached directly to the skin. Figures reproduced with permission from Dagdeviren *et al.* [17].

and produce small voltages; therefore they require the use of specialized equipment such as low-loss coaxial cables and low-noise amplifiers in the measurement chain. Figure 2.18(a) shows how a piezoelectric transducer can be used to measure pressure. The diaphragm in the device moves due to changes in pressure, which in turn generates a voltage. Flexible MEMS fabricated piezoelectric devices are shown in Figures 2.18(b) and (c).

2.4.3.1 Application of Piezoelectric Transducers in Automatic Blood Pressure Measurement

Blood pressure measurements are performed in almost all patients admitted to hospital, and are increasingly being taken automatically rather than manually. The familiar manual (auscultatory) method where the clinician uses a stethoscope to detect a series of what are termed Korotkoff sounds is very accurate, but has the disadvantage of being very sensitive to the exact positioning of the cuff and requiring a quiet acoustic environment. Most automatic clinical machines do not use the auscultatory principle, but rather a more indirect oscillometric method, outlined below. The advantages of this oscillometric method include the features that positioning the cuff correctly is not important and there is no requirement for quiet.

A blood pressure reading consists of two numbers, X over Y. The higher number (X) is the systolic pressure, which corresponds to the pressure exerted by the blood against the artery wall when the heart contracts. The lower number (Y) is the diastolic pressure and represents the pressure against the artery wall between heart beats when the heart is at rest. Typical numbers for healthy adults are 120/80 or lower, where the numbers are measured in mm of mercury (mmHg). In order to measure the blood pressure a cuff is placed around the upper arm and filled with air until the cuff pressure is higher than the systolic blood pressure. The artery in the arm is completely flattened, and no blood is able to pass through. The pressure in

(a)

Voltage from
pressure sensor

(b)

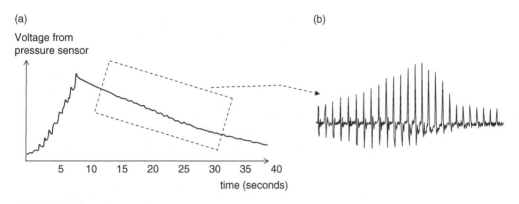

Figure 2.19　(a) Plot of the voltage from a pressure sensor in an automated blood pressure meter. (b) The oscillometric signals that are extracted from the raw voltage measurements and used to produce the systolic and diastolic blood pressure measurements.

the cuff is then lowered by letting some air out. For automatic machines, the cuff is inflated automatically by an electric motor. The oscillometric method works on the principle that when the artery opens during the systolic/diastolic pressure cycle, a small oscillation is superimposed on the pressure inside the cuff. The amplitude of the oscillometric signals varies from person to person, but in general is in the 1 to 3 mmHg range. The frequency of the oscillating signal is ~1 Hz, which is present as a small fluctuation on top of the much larger cuff pressure signal, which has a frequency <0.04 Hz. The output voltage from the piezoelectric pressure sensor in the instrument is shown in Figure 2.19(a). The signal is filtered and amplified (see next chapter) to give the oscillation signal shown in Figure 2.19(b).

The systolic and diastolic blood pressure points are determined by vendor-specific algorithms, which use a combination of the amplitude and the gradient of the signal to find these points. In order to evaluate their choice of points for the systolic and diastolic blood pressure, manufacturers of oscillometric instruments must go through a calibration protocol (ANSI/AAMI SP10), which defines how to validate the calibration for the US market.

2.4.4　Capacitive Transducers

Capacitive sensors are widely used in pressure transducers and accelerometers. This type of sensor is based on changes in the capacitance of a parallel-plate capacitor produced by a change in tissue pressure or displacement. The capacitance can be altered via changes in the overlap area (A) of two metallic plates, or

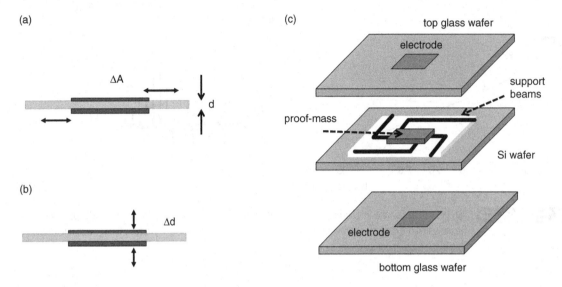

Figure 2.20 (a) Horizontal motion of the two plates produces a change in capacitance via a change in overlap area ΔA. (b) Corresponding situation for vertical motion of the two plates, increasing the separation by Δd. (c) Schematic of the design of a variable capacitive sensor based on MEMS technology.

the distance (d) between the plates, as shown in Figures 2.20(a) and (b), respectively. The capacitance of a parallel-plate capacitor is given by:

$$C = \frac{\varepsilon_r \varepsilon_0 A}{d} \tag{2.27}$$

where ε_r is the relative permittivity of the material between the plates, and ε_0 the permittivity of free space (8.85×10^{-12} Fm^{-1}). A change in overlap area, ΔA, gives a sensitivity $\Delta C/\Delta A$ that is constant:

$$\frac{\Delta C}{\Delta A} = \frac{\varepsilon_r \varepsilon_0}{d} \tag{2.28}$$

whereas a change in distance gives a sensitivity that is non-linear:

$$\frac{\Delta C}{\Delta d} = -\frac{\varepsilon_r \varepsilon_0 A}{d^2} \tag{2.29}$$

Despite the non-linear nature of the second approach, it is much easier for this design to be fabricated in a MEMS structure, and so it is more often used. Figure 2.20(c) shows an example of a microfabricated capacitive sensor.

In addition to the specific example considered in the next section, capacitive pressure sensors can be used for intracranial pressure monitoring and intraocular pressure measurements. They can also be incorporated into the tips of different types of catheters.

2.4.4.1 Application of Capacitive Sensors as Accelerometers in Pacemakers

Capacitive sensors can be used as accelerometers that are integrated into rate-response pacemakers (covered in more detail in section 8.4.1.1). These types of pacemakers are designed to increase the patient's heart rate when they exercise. The output of the accelerometer is used to determine the degree of exercise, and the pacemaker electronics increase the pacing rate applied to the heart accordingly. A simplified diagram of a capacitively based accelerometer is shown in Figure 2.21(a) and (b).

The sensor is usually integrated into a bridge circuit such as that shown in Figure 2.21(c). The circuit can be analyzed very simply:

$$C_1 = \frac{C_0}{1 + \delta y}, C_2 = \frac{C_0}{1 - \delta y} \tag{2.30}$$

where C_0 is the capacitance when $\delta y = 0$. If the change in position $\delta y / y$ is very small, then:

$$V_{out} = \frac{V_{in}}{2} \delta y = -\frac{V_{in}}{2} \frac{ma}{k} \tag{2.31}$$

where m is the mass and k the spring constant of the cantilever, and a is the acceleration. Equation (2.31) shows that the output voltage is proportional to the

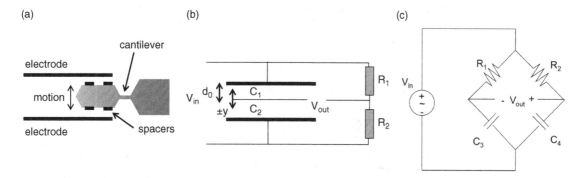

(a)   (b)   (c)

Figure 2.21   Schematic of an accelerometer based on capacitive sensors. (a) The central electrode (light grey), attached to a very thin cantilever moves up and down when the accelerometer is moved or tilted. Separated by air gaps from the central electrode are upper and lower electrodes (black). As the central electrode moves, the capacitance between the electrodes changes (small non-conducting spacers are used to ensure that the electrodes cannot short circuit). The upper and lower electrodes exit the chip at the left, and the central electrode at the right, so that electrical connections can be formed. (b) and (c) The full circuit uses two resistors in a bridge configuration to produce an output voltage that depends upon the sensed acceleration. C_3 and C_4 are equivalent to C_1 and C_2 in (b).

acceleration. Capacitive-based accelerometers are also used to track motion in patients and to provide feedback on exercise regimes and physical therapy, allowing the patient to remain at home rather than have to go to the hospital.

2.4.5 Inductive Transducers: the Linear Voltage Differential Transformer

The final class considered in this section are inductive transducers, which can also be used to measure changes in pressure. The most common device is called a linear voltage differential transformer (LVDT), shown schematically in Figure 2.22. One primary and two secondary coils are connected to an input voltage source in a configuration that results in the output voltage being the difference between the voltages across the two secondary coils. When the core is in its central position, equal and opposite voltages are induced in the two secondary coils, and so the output voltage is zero. When the core is displaced towards the right, the voltage in the secondary coil on the right of Figure 2.22 increases and the voltage in the secondary coil on the left decreases. The resulting output voltage has the same phase as the primary voltage, V_{in}. When the core moves in the other direction, the output voltage has an opposite phase to that of the primary. The phase of the output voltage determines the direction of the displacement (left or right) and the amplitude indicates the amount of displacement.

Linear voltage differential transformers were commonly used in clinical pressure transducers until about a decade ago, but other designs outlined in the previous sections are now preferred. The major issues with the LVDT design are

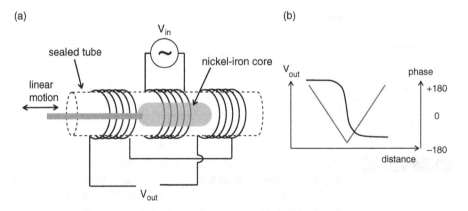

Figure 2.22   (a) Schematic of a linear voltage differential transformer. Displacement of the inner nickel-iron core produces different voltages in the two secondary coils that are situated either side of a primary coil. (b) Plot of the magnitude (grey line) and phase (black line) of the output voltage as a function of the displacement of the core. The phase of the signal with respect to V_{in} changes sign either side of zero displacement.

the requirement for having to use an AC voltage for the measurement, and phase-sensitive detection to determine the sign of the pressure change.

2.5 Chemical Sensors: Example – a Glucose Monitor

Glucose monitors are the most widely used biomedical sensors, with the instrumentation including specialized hospital equipment, portable glucose meters and subcutaneously implanted sensors for continuous monitoring with wireless transmission of data to an external interface and from there to the physician.

2.5.1 Clinical Need for Glucose Monitoring

Glucose monitoring is vital in the management of diabetes. Throughout the world the number of patients with type 1 and type 2 diabetes is increasing rapidly, with many needing regular insulin injections to maintain their glucose levels within healthy limits. Knowledge of blood glucose level is key to perform correct insulin dosing. Over 10 billion glucose measurements per year are performed using portable measuring devices. This translates into an industry with an annual turnover of more than 10 billion US dollars (although the monitors are inexpensive, costing less than 50 dollars, one year's supply of strips, using four strips per day, typically costs between 1000 and 2000 US dollars). In the vast majority of glucose home monitoring systems, the patient pricks their finger with a small needle (lancet) to obtain a drop of blood, places this drop on a disposable 'test strip' that contains chemicals that react with glucose, and the strip is inserted into the glucose meter, which gives a reading in mg/dL (milligrams of glucose per decilitre of blood).

2.5.2 System Requirements for Glucose Monitoring

Laboratory analysis at a hospital is the most accurate method for evaluating glucose levels. It measures levels in blood plasma as opposed to in pure blood as in the case for home glucose meters. Hospital measurements are more accurate due to the presence of trained technicians, controlled temperature and humidity, constant calibration of the machine, a much larger sample of blood (5 ml) compared to home meters (<1 μl), and a longer analysis time.

For home use over fifty different models of portable commercial glucose monitors are currently available. Since glucose monitoring is such an important part of everyday life for patients with diabetes, it is critical that the patients themselves are able to perform the measurement in a simple and accurate fashion. According to the ISO 15197 guidelines a list of system requirements for portable blood glucose monitors includes:

(i) A high specificity to the glucose concentration only. Given that patients may be taking a variety of other therapeutic drugs, it is important that the concentration of these medications in the blood does not influence the estimated glucose level. The readings should also be independent of the haematocrit level, which varies across the menstrual cycle for women, and differs in general among individuals.

(ii) Blood glucose meters must provide results that are within 20% of a laboratory standard 95% of the time. The ISO and FDA have set accuracy criteria to ± 20 mg/dL for glucose levels <100 mg/dL or $\pm 20\%$ for glucose levels >100 mg/dL for at least 95% of the results. The American Diabetes Association has recommended that glucose meters agree to within $\pm 15\%$ of the laboratory method at all glucose concentrations, with a future performance goal of $\pm 5\%$ agreement.

(iii) A high degree of linearity over a wide range of glucose values.

(iv) The device needs to be small, portable and robust given that the patient may have to take readings in a number of different environments, both inside and outside. The readings should be insensitive to altitude, humidity, temperature and other external factors.

(v) The required blood volume should be <1 microlitre, and the needle very thin to minimize patient discomfort, especially given the large number of readings that need to be made over a patient's lifetime.

(vi) The analysis time should be very short to allow a quick response in case of either hypoglycaemia or hyperglycaemia.

(vii) The cost should be less than one US dollar per measurement.

2.5.3   Basic Detection Principles of Glucose Monitoring

The basic detector for most glucose monitoring systems is a chemical transducer that converts the concentration of glucose into a proportional electrical signal via an electrode [18–20]. Most glucometers today use this electrochemical (amperometric) method, in which electrons are exchanged (either directly or indirectly) between an enzymatic sensing system and the electrode [21]. The electron flow in the electrode is proportional to the number of glucose molecules in the blood, providing a linear relationship between glucose concentration and the current produced in the electrode. The two most commonly used enzymatic systems are glucose oxidase (GOx) or glucose-1-dehydrogenase (GDH), both of which are described here. GOx is a large protein, which contains a flavin adenine dinucleotide (FAD) redox (reduction–oxidation) centre deep in its core. Upon reaction with glucose FAD acts initially as an electron acceptor and is reduced to $FADH_2$.

$$\text{Glucose} + \text{GOx(FAD)} + 2H^+ + 2e^- \rightarrow \text{Gluconolactone} + \text{GOx}(FADH_2)$$

In order for the cofactor FAD to be regenerated, $FADH_2$ needs to be oxidized, i.e. to lose electrons. This is usually achieved using synthetic, electron-accepting mediators. In the final step these mediators are re-oxidized at the electrode, with their donated electrons forming the output current. An alternative enzyme system is GDH, which uses osmium complexes as mediators. The general scheme for both GOx and GDH is shown in Figure 2.23, in which gluconolactone is converted to gluconic acid.

The relative advantages and disadvantages of GOx and GDH systems are:

(i) GOx sensors have high specificity and do not cross-react with other sugars in the blood. However, the readings are sensitive to the oxygen concentration in the blood, and so if the SpO_2 levels are low, the blood glucose level is overestimated [22].

(ii) GDH sensors are not affected by oxygen, but there are cross-reactions with other sugars, such as maltose, which can lead to an overestimation of the blood glucose level.

One of the major challenges in accurate glucose measurements is that there may be other chemicals in the blood that also can donate electrons to the electrode and therefore result in an overestimate of the glucose concentration [23]. These include acetaminophen and salicylic acid (both common pain relievers), tetracycline, ascorbic acid, cholesterol and triglycerides. The haematocrit can also have an

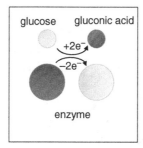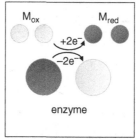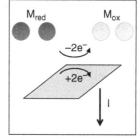

Figure 2.23   Three steps by which electrons are transferred from glucose molecules to the electrode in a glucose meter. M_{ox} and M_{red} represent the oxidized and reduced form of the mediator, respectively. I represents current flow from the electrode.

effect since oxygen from red blood cells can compete with the redox mediator for glucose-derived electrons when GOx is used.

2.5.4   Designing a Portable Device for Glucose Monitoring

The basic scheme for portable glucose monitoring is that the patient pricks their fingertip with a lancet, squeezes out a very small volume of blood, places their fingertip on a disposable test strip, and inserts the strip into a recording device. Figure 2.24 shows a schematic of a disposable test strip, which produces a current proportional to the concentration of glucose in the blood. The strips are produced using MEMS technology outlined earlier in this chapter, and contain a capillary network that sucks up a reproducible amount of blood from the fingertip. Three electrodes are present on the test strip: a reference electrode, a control electrode and a working electrode. The enzyme (either GOx or GDH) and appropriate mediator are deposited on the working electrode by applying them in solution and then evaporating the solvent. A few seconds after the drop of blood (less than half a microlitre is required) has been placed on the strip, the strip is inserted into the recording device,

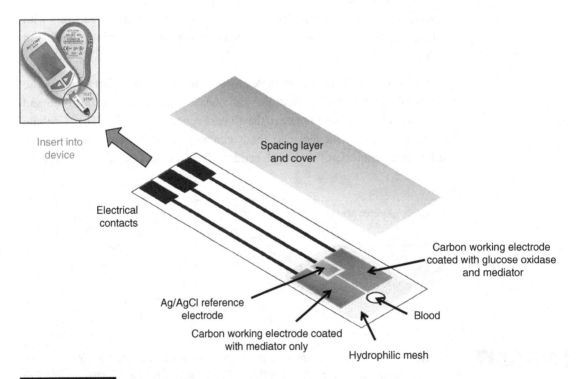

Figure 2.24   Schematic of a test strip used to measure blood glucose levels.

also shown in Figure 2.24. When blood reaches the working electrode the enzyme and mediator dissolve in the blood, and the chemical reactions described in the previous section occur. The current produced from the electrode is in the order of microamps, and the signal must be amplified and filtered as discussed later in section 3.6.2. The major problems associated with finger pricking are that repeated lancing of the finger leads to soreness and associated non-adherence to the required regime of self-testing. This provides motivation for the use of continuous glucose monitoring systems, which are covered in Chapter 8.

2.6   Acoustic Sensors: Example – a Microphone for Hearing Aids

More than 10 million hearing aids are sold annually to patients around the world. Roughly one-third are sold in the United States and Canada, one-third in Europe and one-quarter in Asia and the Pacific Rim. The increase in units sold is approximately 5% annually. Despite these large numbers, it is estimated that only 25 to 35% of people who should be using hearing aids actually are, mainly due to social and economic pressures. There is, therefore, not only tremendous opportunity for economic growth, but also for engineering developments to improve the quality and reduce the size of hearing aids so that they can perform better and be less obtrusive.

2.6.1   Clinical Need for Hearing Aids

Hearing loss affects roughly 10% of the world's population. There are two general classes of hearing loss, sensorineural and conductive, and patients may suffer from both. The most common type of hearing loss is sensorineural which involves damage to the auditory nerve. Sensorineural loss can be compensated by using either a hearing aid that magnifies sound, or a cochlear implant, which stimulates the auditory nerve cells directly. The second cause is conductive loss in which the middle ear is damaged. If the middle ear maintains some degree of function a hearing aid can help by amplification of the signal over the required frequency range. This type of hearing aid is typically a bone anchored hearing aid, implanted into the skull, which transmits sound directly to the hearing nerve. If damage is too severe then a cochlear implant is the clinical treatment plan.

For most patients requiring hearing aids the ability to hear is reduced in the frequency range 500 Hz to 4 kHz. This can be as the result of disease but it also occurs naturally over the human lifespan. Given the aging population in many developed countries, particularly in Germany and Japan, the demand for improved hearing aids is likely to continue for the forseeable future.

2.6.2 Microphone Design for Hearing Aids

The primary sensor in a hearing aid is a microphone, which converts an acoustic signal (sound wave) into a voltage. A directional hearing aid uses two or more microphones to receive sound from multiple directions. This improves the signal-to-noise ratio (SNR) of speech when in a noisy environment, and enhances the quality of speech further when used with digital signal processing, as covered later in Chapter 7.

Most microphone designs are based on electret technology. An electret is a thin (about 12 μm) layer of dielectric, typically Teflon, which is glued on top of a metal back electrode, as shown in Figure 2.25(a), and charged by a corona discharge to a surface potential of approximately 200 V. The electret therefore has a permanent electric charge.

The electret is integrated into the microphone housing as shown in Figure 2.25(b). Sound waves enter the microphone through the sound coupling tube via a thin slit in the outer microphone case, and pass into the front volume of the microphone. The inner wall of the front volume is a thin, flexible polymer diaphragm, which is

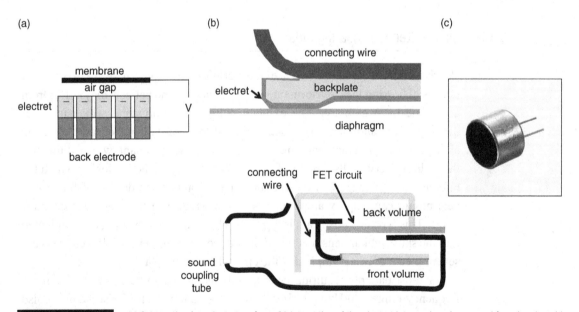

(a)   (b)   (c)

Figure 2.25   (a) Schematic of an electret surface. (b) Integration of the electret into a microphone used for a hearing aid. (c) Photograph of a packaged microphone, which can be as small as a few millimetres in size.

metallized on its lower layer, and electrically connected to the metallic microphone case. A small hole through the diaphragm equalizes the pressures in the front and back volumes. The sound waves cause the diaphragm to move towards the metal backplate with the electret, which induces an opposing electric charge on the diaphragm, effectively forming a parallel plate capacitor. When the diaphragm is displaced towards the electret a small voltage is induced on the backplate, which is electrically isolated from the microphone case. The backplate and the gate of a field emission transistor (FET) amplifier are electrically connected, and the output of the FET represents the transduced acoustic signal. The total size of the microphone can be in the order of 1 to 3 mm, as shown in the photograph in Figure 2.25(c).

A typical frequency response of a microphone is shown in Figure 2.26, which shows that it is relatively flat over the range required for speech intelligibility (500–4,000 Hz). There is a steep decrease in the sensitivity at very low frequencies to reduce the impact of low-frequency noise (which may not be perceptible to the patient, but which would otherwise overload the amplifier). This is achieved using an air gap, which means that low-frequency sounds impinge on both sides of the microphone at essentially the same time (due to the long wavelength), and therefore do not produce any physical displacement of the diaphragm. The larger the gap, the greater the attenuation and also the greater the low-frequency bandwidth over which cancellation takes place.

Noise in the electrical output of the microphone can arise from environmental acoustic noise and noise that is generated within the microphone itself. The internal noise has several potential sources: the microphone electronics, background mechanical motion of the diaphragm or acoustic propagation paths within the microphone. Thompson *et al.* [24] have shown that in the frequency range from 20 Hz to 10 kHz relevant to human speech perception, the total microphone noise is dominated by acoustic noise associated with acoustic flow resistances in the small passages in the microphone. The acoustic resistance has four major sources: (i) flow through the

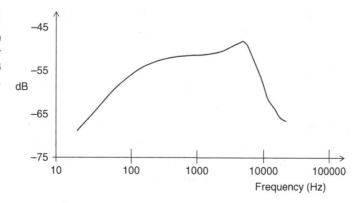

Figure 2.26

Frequency response of the microphone sensitivity for the Knowles EM-3346 microphone.

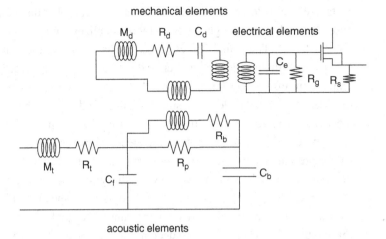

mechanical elements

electrical elements

acoustic elements

Figure 2.27 Electrical circuit equivalent of an electret microphone. M_t is the inertance of the sound entry port, R_t is the resistance of the sound port, R_p is the resistance of the hole in the diaphragm, C_f and C_b are the compliances of the front and back volumes, R_b is the flow resistance between the diaphragm and backplate, M_d is the effective mass of the diaphragm, C_d is the compliance of the diaphragm, which includes the negative compliance from the attractive force between diaphragm and backplate, R_d is the mechanical resistance of the diaphragm, C_e is the diaphragm capacitance, R_g is the bias resistance at the FET gate, and R_s is the FET source resistance. Figure reproduced from Thompson *et al.* [24].

sound inlet tube and the small slot in the outside case; (ii) flow within the front volume; (iii) flow through the hole in the diaphragm; and (iv) flow between the diaphragm and backplate in the back volume. Figure 2.27 shows an electrical circuit analogue of these different effects. For frequencies above 1 kHz the dominant source is flow into the microphone port, whereas for frequencies below 1 kHz it is flow through the hole in the diaphragm.

Microphones for hearing aids can also be produced using MEMS technology, with a total thickness of the device of ~1 mm, similar to that of a thin electret microphone. The main challenges in MEMS design have been to reduce the noise level and power consumption to similar values as for conventional electret technology. Similar to electret-based microphones, MEMS microphones are moving-diaphragm capacitive sensors. They consist of a flexible diaphragm suspended above a fixed backplate, all fabricated on a silicon wafer. This forms a variable capacitor, and a fixed electrical charge is applied between the diaphragm and backplate. The microphone sensor element is constructed using metal deposition and etching processes. The holes in the backplate through which sound waves pass can be less than 10 µm in diameter and the diaphragm thickness can be in the order of 1 µm, with the gap between the diaphragm and the backplate in the order of 1 to 10 µm. A microfabricated preamplifier is connected directly to the backplate. One advantage of the MEMS design is improved stability with respect to change in temperature and humidity since all of the components are fabricated on the same substrate. Another significant advantage of MEMS technology is cost. Many thousands of microphones can be fabricated on a single wafer, whereas electret microphones are typically produced one at a time. MEMS construction also allows submicron control over the most critical dimensions of the microphone.

PROBLEMS

2.1 From a biochemical point of view explain the mechanisms behind the absolute and relative refractory periods for an action potential. If there were no refractory periods, what could happen?

2.2 Using the GHK equation, calculate the resting membrane potentials for cells that are impermeable to all ions except for: (i) sodium, (ii) potassium and (iii) calcium.

2.3 What effect would increasing the membrane permeability of K^+ and Cl^- have on the threshold potential? Would this make it harder or easier to generate an action potential?

2.4 Plot the permeabilities over time of K^+, Na^+ and Ca^{2+} for cardiac pacemaker cells.

2.5 Plot the permeabilities over time of K^+, Na^+ and Ca^{2+} for cardiac ventricular cells.

2.6 Motion of an electrode can cause a disturbance of the Helmholtz double layer surrounding an ECG electrode. Explain what effect this has on the signal recorded by the electrode.

2.7 In a fibre-optic pressure sensor, changes in pressure can be detected by looking at changes in the spectral output as a function of displacement of a flexible diaphragm. In the figure below, waves can undergo multiple reflections, giving two different spectral peaks. Explain how this type of sensor works, and derive an equation to determine the sensitivity of the device. (n is the refractive index of the material, assume a change in dimensions of Δd produced by one wall of the diaphragm moving).

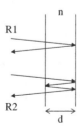
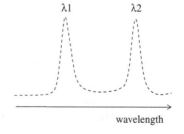

2.8 Think of three physical or physiological factors that could lead to an incorrect pulse oximeter reading. For each factor describe whether the reading would be too high or too low.

2.9 A 10 cm long elastic resistive transducer with a 5 mm diameter has a resistance of 1 kΩ.

(i) Calculate the resistivity of the conductor.
(ii) Calculate the resistance after it has been wrapped around the patient's chest, which has a circumference of 1.2 m, assuming that the cross-section of the conductor does not change.
(iii) Calculate the change in voltage across the transducer assuming that breathing produces a 10% increase in chest circumference and a constant current of 0.5 mA is passed through the conductor.

2.10 In the Wheatstone bridge circuit shown in Figure 2.14 the change in output voltage was analyzed in terms of a change in the value of only one resistor, in this case R_4. The Wheatstone bridge can also be configured such that:

(i) Two-element half-bridge: $R_1 = R_3 = R$, $R_2 = R - \Delta R$, $R_4 = R + \Delta R$.
(ii) Four-element full-bridge: $R_1 = R + \Delta R$, $R_4 = R + \Delta R$, $R_2 = R - \Delta R$, $R_3 = R - \Delta R$.

For each of these three configurations (single element, two-element and four-element) calculate the output voltages as a function of ΔR. Comment on the linearity and sensitivity of the output as a function of ΔR.

2.11 The circuit below shows a Hays bridge, which is used to measure an unknown inductance. Derive the value of the inductance in terms of the other parameters outlined in the figure. Can the circuit be used to measure changes in inductance? Derive appropriate equations.

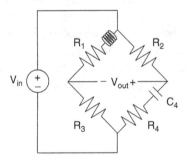

2.12 Calculate the sensitivity of the rotary capacitor shown below with respect to a change in rotation angle. The shaded areas represent copper with the unshaded area of the capacitor having zero conductivity. To what degree is the system linear? Is there a design that can make the system more linear?

2.13 Compare the sensitivities of a three-plate capacitor on the left with a two-plate capacitor on the right.

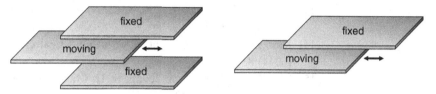

2.14 The device below shows a 'Chevron-shaped' capacitive sensor in which the capacitance changes as the rectangular bar slides from left to right and vice versa. Is this system linear? Can the shape be changed to make it linear? Discuss one advantage of this particular design.

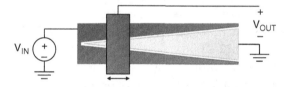

2.15 Design a MEMS device to produce a range of capacitances from 1 pF to 5 pF.

2.16 In capacitance plethysmography a cuff electrode is placed around the limb with a small gap in between, and the change in capacitance between the limb and the electrode measured. Assuming that the limb is a uniform circular cylinder with diameter D, and the spacing s is uniform, calculate the sensitivity of the system in terms of the relationship between the change in capacitance to the change in volume.

2.17 Impedance plethysmography uses a four-band electrode placed around the leg. A constant current is supplied to the two outer electrodes, and the voltage between the two inner electrodes is measured. The ratio of the voltage to current gives the limb impedance. Assume that the electrodes cover a length L, with volume V_0 and tissue resistivity ρ_0. Additional blood with volume V_b and resistivity ρ_b enters the leg and is distributed uniformly in the leg. Derive an expression that relates the change in volume to the measured change in impedance.

2.18 Glucose measurements can be inaccurate due to a high haematocrit level in the blood. By considering separately (a) the electrochemical effects, and (b) the increased blood viscosity at high haematocrit levels, explain whether these two effects separately produce an overestimate or underestimate of the true glucose concentration.

REFERENCES

[1] Saccomandi, P., Schena, E., Oddo, C. M., *et al.* Microfabricated tactile sensors for biomedical applications: a review. *Biosensors* 2014; **4**(4):422–48.

[2] Gad-el-Hak, M. *MEMS: Design and Fabrication.* CRC Press; 2005.

[3] Bhansali, S. & Vasudev, A. *MEMS for Biomedical Applications.* Woodhead Publishing; 2012.

[4] Kim, D. H., Lu, N. S., Ma, R. *et al.* Epidermal electronics. *Science* 2011; **333** (6044):838–43.

[5] Wu, Z. G., Hjort, K. & Jeong, S. H. Microfluidic stretchable radio-frequency devices. *Proc IEEE* 2015; **103**(7):1211–25.

[6] Lorenz, K., Korb, C., Herzog, N. *et al.* Tolerability of 24-hour intraocular pressure monitoring of a pressure-sensitive contact lens. *J Glaucoma* 2013; **22**(4):311–16.

[7] Leonardi, M., Leuenberger, P., Bertrand, D., Bertsch, A. & Renaud, P. First steps toward noninvasive intraocular pressure monitoring with a sensing contact lens. *Invest Ophth Vis Sci* 2004; **45**(9):3113–17.

[8] Mohd-Yasin, F., Nagel, D. J. & Korman, C. E. Noise in MEMS. *Meas Sci Technol* 2010; **21**(1):12–21.

[9] Djuric, Z. Mechanisms of noise sources in microelectromechanical systems. *Microelectron Reliab* 2000; **40**(6):919–32.

[10] Gabrielson, T. B. Mechanical-thermal noise in micromachined acoustic and vibration sensors. *IEEE Trans Electron Dev* 1993; **40**(5):903–9.

[11] Kulah, H., Chae, J., Yazdi, N. & Najafi, K. Noise analysis and characterization of a sigma-delta/capacitive microaccelerometer. *IEEE J Solid-State Circuits* 2006; **41**(2):352–61.

[12] Stenroos, M., Hunold, A. & Haueisen, J. Comparison of three-shell and simplified volume conductor models in magnetoencephalography. *Neuroimage* 2014; **94**:337–48.

[13] Huang, Y., Parra, L. C. & Haufe, S. The New York Head: a precise standardized volume conductor model for EEG source localization and tES targeting. *Neuroimage* 2016; **140**:150–62.

[14] Barlian, A. A., Park, W. T., Mallon, J. R., Rastegar, A. J. & Pruitt, B. L. Review: semiconductor piezoresistance for microsystems. *Proc IEEE* 2009; **97** (3):513–52.

[15] Beebe, D. J., Hsieh, A. S., Denton, D. D. & Radwin, R. G. A silicon force sensor for robotics and medicine. *Sensor Actuat A Phys* 1995; **50**(1–2):55–65.

[16] Wu, C. H., Zorman, C. A. & Mehregany, M. Fabrication and testing of bulk micromachined silicon carbide piezoresistive pressure sensors for high temperature applications. *IEEE Sens J* 2006; **6**(2):316–24.

[17] Dagdeviren, C., Su, Y., Joe, P. *et al.* Conformable amplified lead zirconate titanate sensors with enhanced piezoelectric response for cutaneous pressure monitoring. *Nat Commun* 2014; **5**:4496.

[18] Yoo, E. H. & Lee, S. Y. Glucose biosensors: an overview of use in clinical practice. *Sensors* 2010; **10**(5):4558–76.

[19] McGarraugh, G. The chemistry of commercial continuous glucose monitors. *Diabetes Technol Ther* 2009; **11** (Suppl 1):S17–24.

[20] Oliver, N. S., Toumazou, C., Cass, A. E. & Johnston, D. G. Glucose sensors: a review of current and emerging technology. *Diabet Med* 2009; **26**(3):197–210.

[21] Heller, A. Amperometric biosensors. *Curr Opin Biotechnol* 1996; **7**(1):50–4.

[22] Rebel, A., Rice, M. A. & Fahy, B. G. Accuracy of point-of-care glucose measurements. *J Diabetes Sci Technol* 2012; **6**(2):396–411.

[23] Yoo, E. H. & Lee, S. Y. Glucose biosensors: an overview of use in clinical practice. *Sensors* 2010; **10**(5):4558–76.

[24] Thompson, S. C., LoPresti, J. L., Ring, E. M. *et al.* Noise in miniature microphones. *J Acoust Soc Am* 2002; **111**(2):861–66.

3 Signal Filtering and Amplification

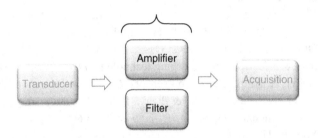

Introduction

After a particular biomedical signal has been detected by the sensor and transduced into an electrical signal, the next steps are to filter and amplify this electrical signal. These two processes may be carried out successively using two different circuits, or the processes can be performed by a single integrated device. The term 'filtering' refers to attenuating any interference signals that occur in a different frequency range to the signals of interest. For example, an EEG signal has frequency components over the range 1 to 50 Hz, but the EEG electrodes also pick up large EOG signals, corresponding to movement of the eye muscles, which typically have frequency components above 100 Hz. In this case, a low-pass filter can be used to reduce the interference from the EOG signal. Amplification is necessary since many biomedical signals have low amplitudes, e.g. EEG signals are in the microvolt and ECG signals in the millivolt ranges. In order to digitize these signals accurately, a process covered in detail in Chapter 4, they must be amplified to the several volts range. Table 3.1 lists some of the characteristics of biomedical signals considered in this book.

Filter designs can be classified as either passive or active. Passive filters are those that do not amplify the signal, whereas active filters both filter and amplify. Examples of both types of circuits are given in this chapter. A very useful tool for analyzing these types of circuits is the Bode plot, which is covered in the following section.

3.1 Frequency-Dependent Circuit Characteristics: Bode Plots

Passive and active filters both use 'lumped element' resistors, inductors and capacitors. When integrated into biosensor design these elements are often

Table 3.1 Characteristics of biomedical signals and interference signals

Sensor	Signal frequencies	Signal amplitude	Interference signal	Interference frequencies	Interference amplitude
ECG electrode	1–50 Hz	1–10 mV	Cardiac motion	0.1–5 Hz	20 mV
			Half-cell potential	DC	220 mV
			Power line pick-up	50/60 Hz	1.5 V
EEG electrode	1–100 Hz	1–10 mV	EOG	100–250 Hz	300 mV
Pulse oximeter	~1 Hz	~10 mV	Scattered light from stationary tissue	DC	1 V
Glucose sensor	DC	~µA	Higher frequency noise	>8 Hz	variable
Hearing aid microphone	1–20 kHz	1 mV	Acoustic flow resistance in microphone	<1 kHz	variable
			Wind signal	<1 kHz	variable

Table 3.2 Properties of inductors (L) and capacitors (C) in terms of voltage (V), current (I) and impedance (Z)

	Time domain		Frequency domain	
Inductor	**Capacitor**		**Inductor**	**Capacitor**
$v(t) = L\frac{di(t)}{dt}$	$v(t) = v_{t=0} + \frac{1}{C}\int_0^t i\,dt$		$\tilde{V} = j\omega L \tilde{I}$	$\tilde{V} = \frac{\tilde{I}}{j\omega C}$
$i(t) = i_{t=0} + \frac{1}{L}\int_0^t V\,dt$	$i(t) = C\frac{dv(t)}{dt}$		$\tilde{I} = \frac{\tilde{V}}{j\omega L}$	$\tilde{I} = j\omega C \tilde{V}$
			$Z = j\omega L$	$Z = -\frac{j}{\omega C} = \frac{1}{j\omega C}$

Figure 3.1

Three simple circuits used to show the principle of the Bode plot.

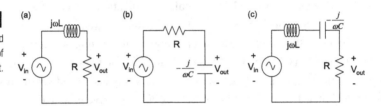

microfabricated using techniques similar to those described at the beginning of Chapter 2. As a brief review Table 3.2 lists the relevant electrical characteristics of capacitors and inductors in both the time domain and frequency domain.

The frequency-dependent *transfer function*, H(jω), of a circuit is defined as the circuit output divided by the circuit input, where the input and output can either be voltages or currents. Circuit analysis tools such as Kirchoff's voltage and current laws, as well as the principles of mesh analysis and superposition, can be used to derive the transfer function in terms of the frequency-dependent impedances of the circuit elements [1].

Having calculated the transfer function, the **Bode plot** is a very simple and quick method for plotting the approximate frequency response of the circuit in terms of its magnitude and phase. It is also very useful in terms of designing filters with specified properties. Two plots are produced: the first is the magnitude of the transfer function measured in decibels (dB) vs. frequency, and the second is the angle of H(jω) vs. frequency.

As an example, the transfer function H(jω) of the series RL circuit in Figure 3.1 (a) can be defined as the output voltage divided by the input voltage:

$$H(j\omega) = \frac{V_{out}(j\omega)}{V_{in}(j\omega)} = \frac{R}{R + j\omega L} = \left(\frac{1}{1 + \frac{j\omega}{R/L}}\right) \tag{3.1}$$

The magnitude Bode plot is calculated on a logarithmic scale:

$$|H(j\omega)|_{dB} = 20\log|H(j\omega)| = 20\log\sqrt{\frac{R^2}{R^2 + \omega^2 L^2}} = 20\log\frac{1}{\sqrt{1 + \left(\omega\frac{L}{R}\right)^2}} \quad (3.2)$$

The quantity R/L is called the break frequency or critical frequency of the circuit. In the Bode plot two separate frequency ranges are considered, one much lower and one much higher than the break frequency, i.e. $\omega \ll$ R/L and $\omega \gg$ R/L, and the corresponding transfer functions are calculated.

$$\text{For } \omega \ll \text{R/L}, |H(j\omega)|_{dB} = 20\log 1 = 0 \text{ dB} \quad (3.3)$$

$$\text{For } \omega \gg \text{R/L}, |H(j\omega)|_{dB} = 20\log\frac{R}{\omega L} \quad (3.4)$$

Equation (3.3) represents a frequency-independent term, whereas equation (3.4) is one where the value of $|H(j\omega)|_{dB}$ decreases as a function of frequency. In order to determine this frequency dependence, consider increases in ω by factors of ten (so-called decades):

$$\begin{aligned}
\omega &= 10(R/L), \ H_{dB} = -20 \ dB \\
\omega &= 100(R/L), \ H_{dB} = -40 \ dB \\
\omega &= 1000(R/L), \ H_{dB} = -60 \ dB
\end{aligned} \quad (3.5)$$

So for frequencies above $\omega =$ R/L a plot of $|H(j\omega)|_{dB}$ is a straight line with a slope of -20 dB/decade.

Adding together the frequency-independent (0 dB) and frequency-dependent terms, the total Bode plot is shown in Figure 3.2.

The Bode plot shows that this circuit acts as a **low-pass filter**, i.e. frequencies below the critical frequency pass through the filter unattenuated, whereas above

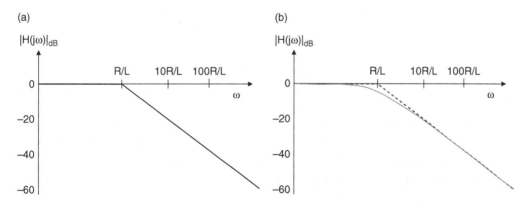

Figure 3.2 (a) Magnitude Bode plot corresponding to the circuit shown in Figure 3.1(a). (b) The exact transfer function is shown in grey, with the bold dotted line representing the Bode straight-line approximation.

the critical frequency they are attenuated, with the higher the frequency the higher the attenuation. Of course, this simple two-line approach is *only* an approximation. For example, there is quite a large error at ω = R/L, where if actual numbers are used:

$$|H(j\omega)|_{dB} = 20\log\left|\frac{1}{1+j1}\right| \approx -3 \ dB \tag{3.6}$$

Therefore one can add in a simple 'error correction' at ω = R/L rad/sec of –3 dB. If the exact expression is evaluated, then the plot is curved rather than straight around ω = R/L, as shown in Figure 3.2(b).

Do these plots make intuitive sense? At very low frequencies the impedance of the inductor is very low, and so almost all of the voltage drop occurs across the resistor, and therefore V_{out}/V_{in} is almost unity, corresponding to 0 dB on a logarithmic scale. As the frequency increases the impedance of the inductor increases and the voltage drop across the resistor decreases, as does the transfer function. At very high frequencies the inductor effectively represents an open circuit, and the transfer function is zero.

The second component of a Bode plot is the phase angle of the transfer function, which reflects the relationship between the phase of the output signal and that of the input signal. For the circuit in Figure 3.1(a) the phase angle is given by:

$$\angle H(j\omega) = \angle\left(\frac{R}{R+j\omega L}\right) = \angle\left(\frac{R^2 - j\omega LR}{R^2 + \omega^2 L^2}\right) = -\tan^{-1}\left(\omega\frac{L}{R}\right) \tag{3.7}$$

In order to see how to plot this, consider three values of ω: ω << R/L, ω = R/L, and ω >> R/L. The corresponding angles of the transfer function are given by:

$$\omega << \frac{R}{10L} : \omega\frac{L}{R} \approx 0, \angle = 0$$
$$\omega = \frac{R}{L} : \omega\frac{L}{R} = 1, \angle = 45°$$
$$\omega >> \frac{10R}{L} : \omega\frac{L}{R} \approx \infty, \angle = 90° \tag{3.8}$$

The corresponding Bode phase angle plot is shown Figure 3.3(a).

Once again, the straight-line segments are only an approximation to the exact solution, as shown in Figure 3.3(b). Nevertheless, the approximation gives a good indication of the circuit behaviour.

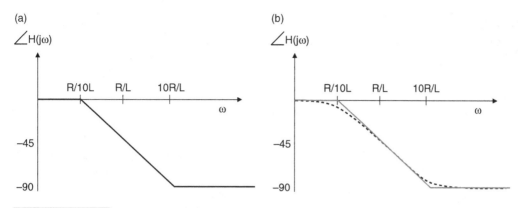

Figure 3.3 (a) Bode plot of the phase angle of the transfer function of the RL circuit. (b) The exact numerical solution shown as the curved grey line, with the Bode approximation as the dotted line.

One could characterize an alternative frequency-dependent impedance of the circuit by defining a different transfer function, in which case:

$$H(j\omega) = \frac{V_{in}(j\omega)}{I_{in}(j\omega)} = Z(j\omega) \tag{3.9}$$

Again, the transfer function consists of one frequency-independent and one frequency-dependent term. The frequency-independent component is given by:

$$|Z(j\omega)|_{dB} = 20\log(R/L) \tag{3.10}$$

The frequency-dependent term is given by:

$$|Z(j\omega)|_{dB} = 20\log\sqrt{1^2 + \left(\frac{\omega}{R/L}\right)^2} \tag{3.11}$$

The phase term is given by:

$$\angle Z(j\omega) = \angle\left(1 + \frac{j\omega}{R/L}\right) = \tan^{-1}\left(\omega\frac{L}{R}\right) \tag{3.12}$$

Figure 3.4 shows the corresponding magnitude and phase Bode plots for the circuit impedance. The plots show that at low frequencies the impedance is essentially that of the resistor, whereas at high frequencies it is that of the inductor, as expected.

As a second example the voltage transfer function of the RC circuit in Figure 3.1 (b) is easily derived by voltage division as:

Figure 3.4

Magnitude and phase
Bode plots of the
impedance of the circuit in
Figure 3.1(a).

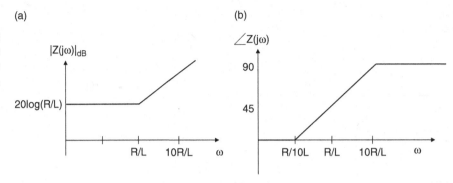

(a)

(b)

The magnitude Bode plot is derived from the expression:

$$H(j\omega) = \frac{V_{out}(\omega)}{V_{in}(\omega)} = \frac{1/j\omega C}{R + 1/j\omega C} = \frac{1}{j\omega CR + 1} = \frac{1}{1 + \frac{j\omega}{(1/RC)}} \tag{3.13}$$

The magnitude Bode plot is derived from the expression:

$$|H(j\omega)|_{dB} = 20\log\left(\frac{1}{1 + j\omega CR}\right) = 20\log\left(\frac{1}{\sqrt{1 + \left(\frac{\omega}{(1/RC)}\right)^2}}\right) \tag{3.14}$$

The break frequency is given by 1/RC radians per second. Applying the same two conditions of $\omega \ll (1/RC)$ and $\omega \gg (1/RC)$ gives the magnitude plot shown in Figure 3.5(a). This circuit also acts as a low-pass filter. The phase of the transfer function is given by:

$$\angle H(j\omega) = \angle\left(\frac{1}{1 + j\omega RC}\right) = -\angle(1 + j\omega RC) = -\tan^{-1}\left(\omega\frac{1}{RC}\right) \tag{3.15}$$

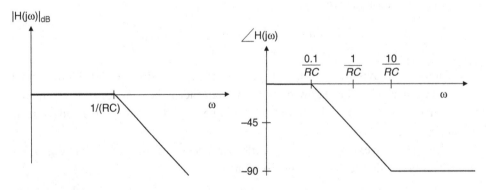

Figure 3.5   (a) Magnitude and (b) phase Bode plots of the voltage transfer function of the RC circuit shown in Figure 3.1(b).

The phase angle Bode plot is shown in Figure 3.5(b).

The final example in Figure 3.1(c) shows a circuit that can be described as a second-order filter. The voltage-transfer function can be derived from voltage division and is given by:

$$H(j\omega) = \frac{V_{out}(\omega)}{V_{in}(\omega)} = \frac{R}{R + \frac{1}{j\omega C} + j\omega L} = \frac{j\omega RC}{j\omega RC + 1 - \omega^2 LC} \tag{3.16}$$

In this case, the plot is most easily visualized by substituting in values: for example $R = 100\ \Omega$, $C = 20\ \mu F$, $L = 50\ mH$. Using these values the transfer function can be expressed as:

$$H(j\omega) = \frac{j\omega 2000}{-\omega^2 + j\omega 2000 + 10^6} = \frac{j\omega 2000}{(j\omega + 1000)^2} = \frac{j\omega(1/500)}{(1 + j\omega/1000)^2} \tag{3.17}$$

Applying the same analysis principles as previously the Bode plots shown in Figure 3.6 can be derived (see Problem 3.2). This magnitude plot shows that the circuit acts as a band-pass filter, allowing a certain band of frequencies to pass through, while strongly attenuating higher and lower frequencies. The phase plot indicates that the circuit looks inductive at low frequencies and capacitive at higher frequencies.

Bode plots can be calculated for any arbitrary transfer function. If the form of the transfer function is given by:

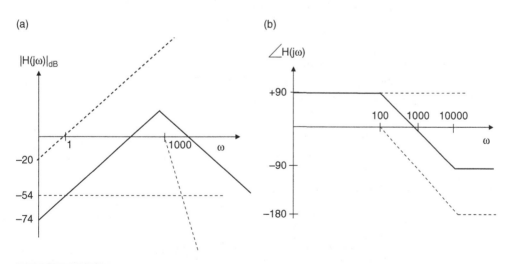

Figure 3.6   (a) Magnitude and (b) phase angle plots of the voltage-transfer function of the circuit shown in Figure 3.1(c). Individual components from equation (3.17) are shown as dotted lines, with the solid line representing the sum of the individual components, i.e. the overall transfer function.

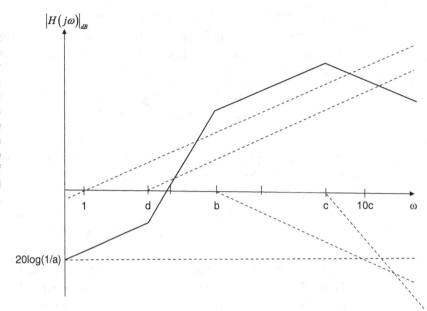

Then the magnitude plot is given by the summation of individual terms:

$$H(j\omega) = \frac{\dfrac{j\omega}{a}\left(1+\dfrac{j\omega}{d}\right)}{(1+j\omega/b)(1+j\omega/c)^2} \tag{3.18}$$

$$|H(j\omega)|_{dB} = 20\Big(\log\omega + \log(1/a) + \log(\omega/d) - \log(\omega/b) - \log(\omega/c)^2\Big) \tag{3.19}$$

Figure 3.7 shows the individual components and total magnitude Bode plot for this general function. Similarly, the angle plot can also be broken down into individual components:

$$\angle H(j\omega) = \angle j\omega + \angle(1/a) + \angle(1+j\omega/d) - \angle(1+j\omega/d) - \angle(1+j\omega/c)^2 \tag{3.20}$$

with the phase angles given by:

$$\begin{aligned}
&\angle j\omega = 90 \\
&\angle(1/a) = 0 \\
&\angle(1+j\omega/d) = 0^0{}_{\omega<0.1d}, 45^0{}_{\omega=d}, 90^0{}_{\omega>10d} \\
&-\angle(1+j\omega/b) = 0^0{}_{\omega<0.1b}, -45^0{}_{\omega=b}, -90^0{}_{\omega>10b} \\
&-\angle(1+j\omega/c)^2 = 0^0{}_{\omega<0.1c}, -90^0{}_{\omega=c}, -180^0{}_{\omega>10c}
\end{aligned} \tag{3.21}$$

Figure 3.8 shows the individual components and the total phase Bode plot.

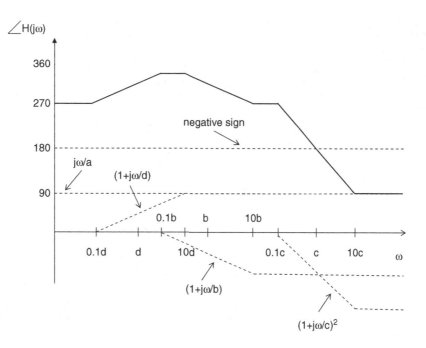

Figure 3.8

Bode phase plot corresponding to the generalized transfer function of equation (3.18). Individual components are shown as dotted lines, with the overall transfer function given by the bold solid line representing the summation of these individual components.

Bode plots are very useful to design and analyze circuits used as filters, which are covered in the next sections.

3.2 Passive Filter Design

As discussed at the end of Chapter 2, the biosignals produced from a range of different sensors can also contain significant interference from other sources. If these interfering sources have a frequency spectrum that does not overlap with that of the signals of interest, then they can be removed using a frequency-selective filter. A general outline for designing passive and active filters is shown in Figure 3.9 [2].

The first step in the design process is to determine the type of filter, i.e. low-pass, high-pass, band-pass or band-stop/notch. Next, the required order of the filter should be determined, with a higher order providing a greater degree of filtering but requiring more electrical components in the design, with greater associated losses. The third step is to consider (for second or higher order filters) whether having a very flat response in the passband but a relatively shallow transition (for example a Butterworth filter) or accepting some ripple in the passband with the advantage of a steeper transition (one example being a Chebyshev filter) is more appropriate for the particular application. Although only Butterworth and

Figure 3.9

General scheme showing
the different design
decisions in determining
which filter design should
be used. In the upper
three panels the vertical
scale represents the
magnitude of the transfer
function, and the
horizontal axis the
frequency.

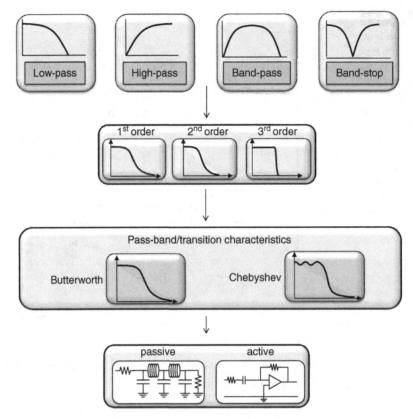

Figure 3.9 General scheme showing the different design decisions in determining which filter design should be used. In the upper three panels the vertical scale represents the magnitude of the transfer function, and the horizontal axis the frequency.

Chebyshev are considered here in detail, there are also a number of other geometries, such as elliptical and Bessel, which can also be used. The final decision is whether the filter should be passive or active. As covered later in this chapter, the highest SNR is achieved when amplification is used before, or integrated with, a filter (the latter case constituting an active filter). This approach can be used unless the interfering signal is so much greater than the signals of interest that it would saturate the input of the amplifier, in which case a passive filter is used before the amplification stage.

3.2.1 First-Order Low-Pass and High-Pass Filters

Many different configurations of first-order (where the order refers to the number of reactive components, i.e. capacitors or inductors) low-pass and high-pass filters can be designed using resistors, capacitors and inductors: some examples are shown in Figure 3.10. Bode plots for these types of circuit have already been

Figure 3.10

Four possible geometries for first-order low-pass filters (LPF) and high-pass filters (HPF).

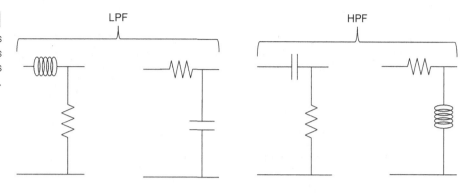

LPF   HPF

Figure 3.11

Different configurations of (a) second-order low-pass and (b) band-pass filters.

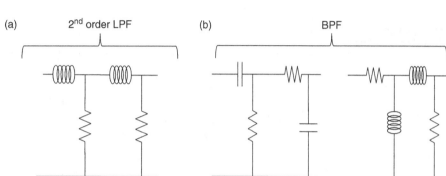

(a)   2ⁿᵈ order LPF   (b)   BPF

derived. As shown previously, the attenuation produced by these first-order filters increases at a rate of 20 dB per frequency decade.

3.2.2 Higher Order High-Pass, Low-Pass, Band-Pass and Band-Stop Filters

In order to increase the frequency-dependent attenuation to more than 20 dB per decade, first order filters can be combined together to form higher order filters, as shown in Figure 3.11(a). The Bode plot is simply the sum of the individual first-order components. By combining high-pass and low-pass circuits in series, with an appropriate choice of the respective cut-off frequencies, second-order band-pass filters can be produced, as shown in Figure 3.11(b). For band-stop circuits, the individual high-pass and low-pass circuits can be combined in parallel, although in practice band-stop circuits are more commonly formed from a parallel resonant circuit covered in section 3.2.3.

Although cascading first-order filters appears to be a very simple and intuitive method of designing higher order filters, in practice each successive RL or RC module is 'loaded' by the previous one, and if one considers the real response (rather than the idealized Bode plot) the region around the cut-off frequency is

(a)   (b)

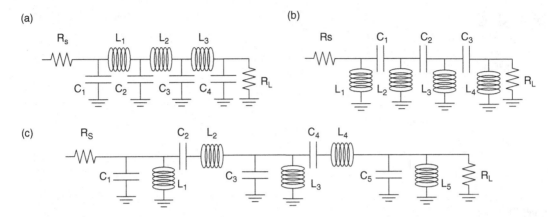

(c)

Figure 3.12   General design of passive multi-pole filters. (a) Low-pass, (b) high-pass and (c) band-pass. The component values are determined by the class of filter, e.g. Butterworth or Chebyshev, and the required cut-off frequency.

attenuated very slowly: only at frequencies much higher than the cut-off frequency does the 20 N dB (where N is the order of the filter) per decade fall-off occur.

Instead, practical higher order filters typically consist of several inductors and capacitors as shown in Figure 3.12. The particular values of the capacitors and inductors determine the frequency response of the filter, as discussed below. For simplicity, it is assumed that the source and load resistances, R_s and R_L respectively, are the same. However, it is also possible to design these types of filters for arbitrary input and load impedances.

Two of the most commonly used filters are termed Butterworth and Chebyshev. A Butterworth filter has a maximally flat response in the passband, whereas the Chebyshev has an equiripple response in the passband, but minimizes the difference between the ideal 'top-hat' frequency characteristics and the actual response over the entire range of the filter. As seen in Figure 3.13, the Chebyshev filter has a sharper roll-off than the Butterworth. The choice of filter very much depends upon the particular application.

The transfer function for a Butterworth filter is given by:

$$\frac{V_{out}(j\omega)}{V_{in}(j\omega)} = \frac{1}{\sqrt{\left(1 + \left(\frac{\omega}{\omega_c}\right)^{2N}\right)}} \qquad (3.22)$$

where ω_c is the cut-off (break) frequency, and N is the order of the filter. The component values for a given order can be calculated using normalized Butterworth coefficients shown up to sixth-order in Table 3.3, with values given both for pi- as well as T-networks.

Table 3.3 Normalized Butterworth coefficients for filters of order N for a π-network (upper number) and T-network (lower number) for a source resistance of 1 Ω, load resistance of 1 Ω and –3 dB cut-off frequency of 1 radian/second

N		Z1	Z2	Z3	Z4	Z5
2	π	1.414	1.414			
	T	1.414	0.707			
3	π	1.000	2.000	1.000		
	T	1.500	1.333	0.500		
4	π	0.765	1.848	1.848	0.765	
	T	1.531	1.577	1.082	0.383	
5	π	0.618	1.618	2.000	1.618	0.618
	T	1.545	1.694	1.382	0.894	0.309

Figure 3.13

Comparison of the general characteristics of (a) Butterworth and (b) Chebyshev filters in terms of transition width and passband ripple.

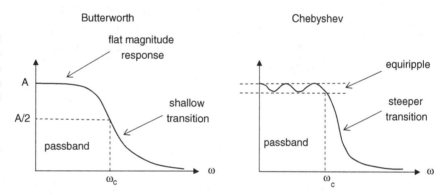

Table 3.3 gives normalized values, and in order to calculate the design for a particular cut-off frequency ω_c the scaling rules for a low-pass filter are given by:

$$L_N(component) = \frac{L_N(normalized)}{\omega_c}, \; C_N(component) = \frac{C_N(normalized)}{\omega_c}$$

$$(3.23)$$

Similarly, to scale for different input resistances, $R_{source} = R_{load} = R$:

$$L_N(component) = L_N(normalized) \text{x } R, \; C_N(component) = \frac{C_N(normalized)}{R}$$

$$(3.24)$$

Example

Design a third-order low-pass Butterworth filter with a 3 dB cut-off frequency of 100 MHz, and 50 Ω source and load impedances.

Solution

There are two possible geometries, a π-network or a T-network with the geometries shown below.

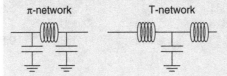

π-network   T-network

Using the values of Butterworth coefficients (1 2 1) given in Table 3.3 for a pi-network of third order, and the scaling formulae given in equations (3.23) and (3.24), the value of the capacitors and inductors is given by:

$$C = \frac{1}{2\pi(10^8)50}, \quad L = \frac{2(50)}{2\pi(10^8)}$$

These expressions give a capacitor value of ~32 pF and an inductor of ~160 nH.

Using the values of Butterworth coefficients (1.5 1.333 0.5) for a T-network of third order, and the appropriate scaling formulae gives a capacitor value of 64 pF and inductors of 80 nH.

Another very commonly used filter is the Chebyshev topology, which is often used when a very sharp transition is required. The transfer function is given by:

$$\frac{V_{out}(j\omega)}{V_{in}(j\omega)} = \frac{1}{\sqrt{1 + \varepsilon^2 C_N^2\left(\frac{\omega}{\omega_c}\right)}} \tag{3.25}$$

where ε is the ripple, C_N are the Chebyshev polynomials of the first kind of order N, and ω_c is the cut-off frequency, which in this case is defined as the highest frequency at which the filter 'gain' falls to the same value as the ripple, as shown in Figure 3.13. Table 3.4 shows the normalized Chebyshev coefficients required for the design of these filters.

Table 3.4 Normalized Chebyshev coefficients for filters of order N

N	Z_1	Z_2	Z_3	Z_4	Z_5	Z_6	Z_7
2	1.403	0.707	0.504				
3	1.596	1.097	1.596				
4	1.670	1.193	2.367	0.842	0.050		
5	1.706	1.230	2.541	1.230	1.706		
6	1.725	1.248	2.606	1.314	2.476	0.870	0.504

Figure 3.14

Two configurations of a parallel RLC circuit that can be used as a filter. (a) Geometry that acts as a band-pass filter. (b) Circuit that serves as a band-stop or notch filter.

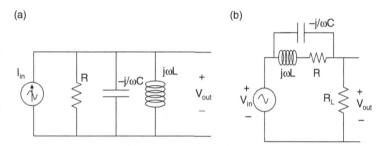

3.2.3 Resonant Circuits as Filters

Another class of circuit that is extensively used in filters (as well as other applications covered elsewhere in this book) is termed a resonant circuit. A resonant circuit is one in which the output (or transfer function) has a maximum value at a particular frequency, and decreases rapidly either side of this frequency, i.e. it acts as a sharp band-pass filter. Circuits tuned to a specific frequency are used in many areas of data transmission including radio, television, satellite and global positioning systems (GPS). Resonant circuits are also used extensively in wireless medical applications, which are covered in Chapter 8, since they can be used to power devices wirelessly or to transfer data wirelessly between a transmitter and a receiver.

There are two variants of resonant circuits, namely parallel and series. The band-pass nature of the series RLC circuit in Figure 3.1(c) has already been demonstrated in Figure 3.6. Figure 3.14 shows two configurations of a parallel RLC circuit.

For the configuration shown in Figure 3.14(a) the output voltage is given by:

Figure 3.15

(a) Plot of the magnitude of the transfer function of the parallel RLC circuit shown in Figure 3.14(a). (b) Corresponding phase plot of the transfer function.

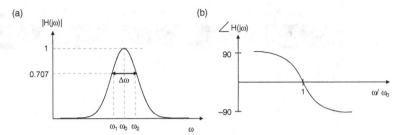

$$V_{out}(j\omega) = \frac{I_{in}(j\omega)}{\sqrt{\frac{1}{R^2} + \left(\omega C - \frac{1}{\omega L}\right)^2}} \tag{3.26}$$

The output voltage has a maximum value at a resonance frequency, ω_0, given by:

$$\omega_0 C = \frac{1}{\omega_0 L} \Rightarrow \omega_0 = \frac{1}{\sqrt{LC}} \tag{3.27}$$

At the resonant frequency V_{out} is equal to $I_{in}R$. The magnitude and phase plots for the transfer function $H(j\omega) = V_{out}(j\omega)/I_{in}(j\omega)$ are shown in Figure 3.15. The phase plot shows that the circuit looks inductive at low frequency, capacitive at high frequency and purely resistive at the resonance frequency.

The bandwidth ($\Delta\omega$) of the band-pass region is defined as the difference between frequencies ω_1 and ω_2, which correspond to the frequencies at which the value of the transfer function has decreased to 0.707 times its maximum value. These two frequencies are given by:

$$\omega_1 = -\frac{1}{2RC} + \sqrt{\left(\frac{1}{2RC}\right)^2 + \frac{1}{LC}}, \quad \omega_2 = +\frac{1}{2RC} + \sqrt{\left(\frac{1}{2RC}\right)^2 + \frac{1}{LC}} \tag{3.28}$$

The bandwidth is given by:

$$\Delta\omega = \omega_2 - \omega_1 = \frac{1}{RC} \tag{3.29}$$

The quality (Q) factor of the circuit value is defined as:

$$Q_p = \frac{\omega_0}{BW} = \frac{R_p}{\omega_0 L} = R_p\sqrt{\frac{C}{L}} \tag{3.30}$$

A parallel RLC circuit can also be used as a band-stop filter in a configuration such as shown in Figure 3.12(b) (see Problem 3.8).

3.3 Operational Amplifiers

Operational amplifiers, universally referred to as op-amps, are the most commonly used electronic devices used to amplify small voltages in biomedical devices. They consist of networks of many transistors, and are active devices in the sense that they require a DC voltage to drive them [3, 4], as shown in Figure 3.16.

The op-amp amplification factor, referred to as gain, is usually specified in terms of the **gain–bandwidth product (GBWP)**, which is the product of the amplifier's bandwidth and the gain over that bandwidth. If the GBWP of an op-amp is specified to be 1 MHz, it means that it will work up to 1 MHz with unit gain without excessively distorting the signal. The same device, when incorporated into a circuit with a gain of 10, only works up to a frequency of 100 kHz.

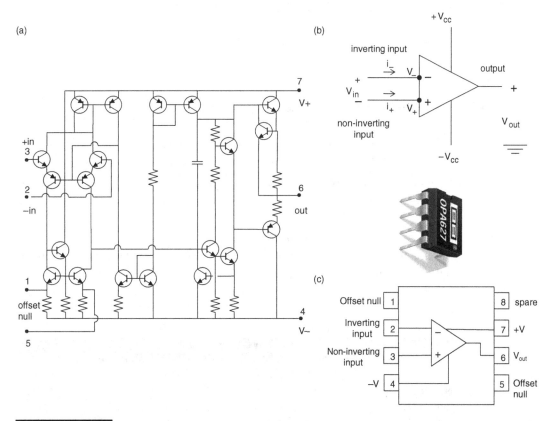

Figure 3.16 (a) Typical transistor layout of an op-amp. (b) Schematic circuit of an op-amp, where V_{cc} represents the collector supply voltage that powers the op-amp. (c) Photograph of a single op-amp chip, and the layout of an eight-pin op-amp.

3.3.1 Circuit Analysis Rules for Op-Amps

The input resistance of an op-amp is very high, approximately 1 MΩ, and so with typical driving voltages of between 5 and 15 volts the input currents i_+ (to the non-inverting terminal) and i_- (to the inverting terminal) are between 5 and 15 picoamps. Therefore a simplifying assumption used extensively in op-amp circuit analysis is that there is no current flow into the device, i.e.

$$i_+ = i_- = 0 \tag{3.31}$$

The gain of the op-amp is defined as the ratio of the output voltage, V_{out}, to the input voltage, (V_{out}/V_{in}) shown in Figure 3.16(b). The maximum output voltage, known as the saturation voltage, is equal to the driving voltage, V_{cc}. Since the typical gain of an op-amp is 10^6, this means that the maximum voltage that can be amplified is ~5 to 15 μV. This is a very small value, and a second approximation used in circuit analysis is that this can be ignored, i.e.

$$V_+ - V_- = 0 \tag{3.32}$$

3.3.2 Single Op-Amp Configurations

There are a number of different op-amp configurations that are commonly used in biomedical instrumentation. Table 3.5 summarizes some of these configurations, together with their properties. The following sections provide basic analyses of these circuits.

3.3.2.1 Inverting and Non-Inverting Amplifiers

The circuit for an inverting amplifier is shown in Figure 3.17. The positive terminal of the input source (V_{in}) is connected to the negative (inverting) terminal of the op-

Table 3.5 Properties of different single op-amp configurations

Configuration	Properties
Inverting	Active gain and the output has opposite polarity to the input
Non-inverting	Active gain and the output has the same polarity as the input
Integrating	Amplifies the time integral of the input signal
Differentiating	Amplifies the time differential of the input signal
Buffer	Unity gain: buffers the input with respect to different loads
Differential	Amplifies the difference between two input signals

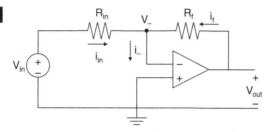

Figure 3.17

Circuit diagram for an
inverting op-amp.

amp, and the positive terminal (non-inverting) of the op-amp is connected to ground. For simplicity, the driving voltages $+/-V_{cc}$ are omitted from the circuit diagrams, but are of course always present.

Writing down expressions for the currents in Figure 3.17:

$$i_{in} = \frac{V_{in} - V_-}{R_{in}}, i_f = \frac{V_{out} - V_-}{R_{in}} \tag{3.33}$$

$V_+ = 0$ as it is connected to ground, and as a result $V_- = 0$ from equation (3.30). Therefore:

$$i_{in} = \frac{V_{in}}{R_{in}}, i_f = \frac{V_{out}}{R_{in}} \tag{3.34}$$

Since $i_- = 0$ from equation (3.29), $i_1 = -i_f$, and the value of V_{out} is given by:

$$\boxed{V_{out} = -\frac{R_f}{R_{in}} V_{in}} \tag{3.35}$$

The result shows that the gain of the amplifier is dictated *only* by the relative values of the two resistors, with a maximum value of $+/-V_{cc}$. The polarity of the output is inverted with respect to the input.

The circuit for a non-inverting amplifier is shown in Figure 3.18. Now the positive terminal of the input source is connected to the positive (non-inverting) terminal of the op-amp.

As before, $V_+ = V_{in} = V_-$ and so the current $i_1 = V_{in}/R_1$. Since $i_- = 0$ then $i_f = i_1$. Applying Kirchoff's voltage law around the whole loop:

$$\boxed{V_{out} = V_- + R_f i_f = V_{in} + R_f \frac{V_{in}}{R_{in}} = \left(1 + \frac{R_f}{R_{in}}\right) V_{in}} \tag{3.36}$$

For this configuration, the output of the amplifier has the same polarity as V_{in}. For values of $R_f \gg R_1$ the gain factor is the same as for the inverting op-amp configuration.

Figure 3.18

Circuit diagram for a non-inverting op-amp.

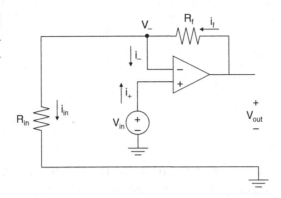

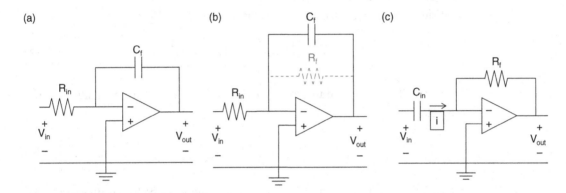

Figure 3.19   (a) A simple integrating amplifier circuit; (b) a practical realization of an integrating circuit in which a large resistor is added in parallel with the capacitor in the feedback loop; and (c) a differentiating amplifier circuit.

3.3.2.2   Integrating and Differentiating Amplifiers

The inverting and non-inverting op-amp circuits contain only resistive elements. By introducing reactive (capacitive and/or inductive) elements, different properties can be introduced into op-amp circuits. Figure 3.19(a) shows a simple realization of an integrating amplifier, which incorporates a capacitor into a feedback loop between the output of, and input to, the op-amp. Since no current enters the op-amp, the current through the resistor R_{in} must be equal to that through the capacitor giving:

$$\frac{V_{in}}{R_{in}} = -C_f \frac{dV_{out}}{dt} \tag{3.37}$$

This means that the output voltage corresponds to the *time integral* of the input voltage.

$$V_{out} = -\frac{1}{R_{in}C_f}\int V_{in}\,dt + \text{constant} \qquad (3.38)$$

As will be discussed in the later section on active filters, this integrating circuit is very sensitive to small offsets in the DC driving voltage and bias current. In order to reduce this sensitivity a large resistor (typically several tens of mega-ohms) is placed in parallel with the capacitor, as shown in Figure 3.19(b). This reduces the cut-off frequency (1/RC) and so the effects of DC variations are reduced.

By moving the capacitor from the feedback to the input arm of the op-amp as shown in Figure 3.19(c), the circuit can be converted from an integrator to a differentiator. Using the usual assumptions that $V_+ = V_- = 0$, and $i_+ = i_- = 0$:

$$i = C\frac{dV_{in}}{dt}$$

Using Kirchoff's voltage law around the outside of the circuit one can derive the expression for V_{out}:

$$V_{out} = -R_f C_{in}\frac{dV_{in}}{dt} \qquad (3.39)$$

3.3.2.3 Buffer Amplifiers

The buffer amplifier is a special case of an op-amp circuit in which $R_f = R_{in} = 0$, as shown in Figure 3.20.

One might well ask what is the point of such a circuit! It is designed to buffer the output voltage of a driving source from changes in the input resistance of a connected load. As an example, in the circuit shown in Figure 3.21(a) suppose that a driving voltage of +5 volts is required to drive a device with an input resistance R_L. A 10-volt power supply is used with the driving voltage provided across half of the 2 kΩ output resistance of the source. If $R_L \gg 1$ kΩ, then the voltage across the device is indeed very close to +5 volts: for example, if $R_L = 100$ kΩ then the driving voltage is 4.995 volts. However, if R_L is comparable to or less than the source's output resistance then the driving voltage drops and the device may not operate correctly: for example, if $R_L = 1000\ \Omega$, then the driving

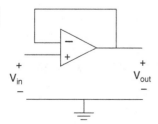

$+$
V_{in}
$-$

$+$
V_{out}
$-$

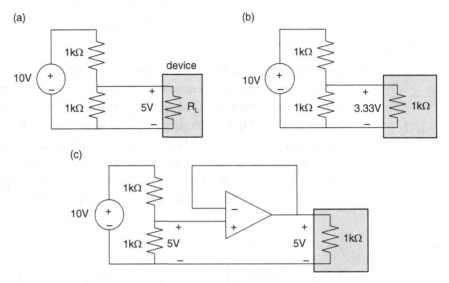

Figure 3.21

Example of the operation
of a buffer op-amp to
buffer the effect of the
input resistance of
a device from the voltage
source. (a) Shows the set
up with the device
connected directly to the
output of the voltage
divider, assuming that R_L
$\gg 1$ kΩ. In case (b) the
voltage across R_L drops
below 5 volts. (c) Adding
a buffer amplifier between
the source and device
restores the driving
voltage to 5 volts.

voltage drops to 3.33 volts, as shown in Figure 3.21(b). In contrast, if a buffer amplifier is placed between the source and device, as shown in Figure 3.21(c), then the input voltage to the device is 5 volts *irrespective of the input resistance* of the device, thus **buffering** the source from the device.

This buffering ability is also useful for applications in which the impedance of the sensor/detector may change over time. One example occurs for biopotentials measured on the scalp (EEG) or chest (ECG) using electrodes, in which the skin–electrode impedance can change over time, thus potentially altering the detected signal independent of any physiological changes. As seen later in this chapter in section 3.3.3, buffer amplifiers are intrinsic components in an instrumentation amplifier, which is very widely used in biomedical devices.

3.3.2.4   Summing and Differential Amplifiers

As their names suggest summing and differential amplifiers have outputs that represent the sum of, or difference between, two input voltages, respectively. Figure 3.22(a) shows a circuit for a summing amplifier. Using the analysis presented for the inverting op-amp, the output voltage is given by:

$$V_{out} = -R_f \left(\frac{V_1}{R_1} + \frac{V_2}{R_2} \right) \tag{3.40}$$

In the case that $R_1 = R_2 = R$, then V_{out} is proportional to $V_1 + V_2$:

Figure 3.22

Circuit diagram for (a)
a summing amplifier and
(b) a differential amplifier.

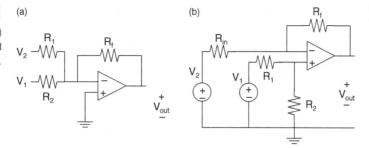

$$V_{out} = -\frac{R_f}{R}(V_1 + V_2) \tag{3.41}$$

Figure 3.22(b) shows the basic circuit for a differential amplifier. Many techniques can be used to analyze this circuit: here the principle of superposition is used, in which the output from each voltage source is considered separately, with the other source short circuited. First considering the effect of V_1 alone, the circuit becomes that shown in Figure 3.23.

By voltage division:

$$V_+ = \frac{R_2}{R_2 + R_1}V_1 \tag{3.42}$$

Then recognizing that this is the non-inverting configuration covered previously, with input V_+:

$$V'_{out} = \left(1 + \frac{R_f}{R_{in}}\right)\left(\frac{R_2}{R_2 + R_1}\right)V_1 = AV_1 \tag{3.43}$$

In the second superposition step, replacing the source V_1 with a short circuit gives Figure 3.24.

This is a simple inverting configuration with the output given by:

$$V''_{out} = -\frac{R_f}{R_{in}}V_2 = -BV_2 \tag{3.44}$$

Therefore the total output of the circuit with both V_1 and V_2 present is given by:

$$\boxed{V_{out} = AV_1 - BV_2 = \left(1 + \frac{R_f}{R_{in}}\right)\left(\frac{R_2}{R_1 + R_2}\right)V_1 - \frac{R_f}{R_{in}}V_2} \tag{3.45}$$

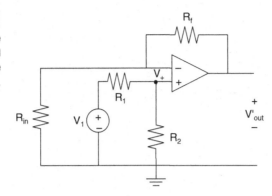

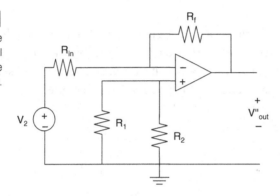

If $R_1 = R_2 = R_f = R_{in}$ then $A = B = 1$, and the result is a differential amplifier with unity gain. The values of the resistors can be chosen to give values of A and B greater than one if a gain factor is required for the differential output.

Many biomedical measurements are intrinsically differential in nature and therefore use some form of differential amplifier. For example, bipolar ECG leads, covered in detail in Chapter 5, measure the potential difference between electrodes placed at different positions on the body. As shown in section 2.3.3, the signal from each electrode also contains a large DC component (hundreds of millivolts), which would saturate the op-amp. Electrode measurements also contain significant components of 50/60 Hz noise picked up by the electrode leads from nearby power sources. In an ideal case, these two types of interfering signals are identical in each electrode, and therefore would be cancelled out by feeding the inputs into the differential amplifier shown in Figure 3.22(b). One can quantify the degree of cancellation via a quantity known as the common mode rejection ratio (CMRR), defined as:

$$CMRR = \frac{G_d}{G_c} \tag{3.46}$$

where G_d is the differential gain (how much the desired difference voltage is amplified), and G_c is the common mode gain (how much the undesired signal components are amplified). The value of CMRR is usually given in dB ($CMRR_{dB} = 20 \log G_d/G_c$), so a CMRR of 60 dB corresponds to a ratio of 1000 in differential:common mode gain. In reality, a single differential op-amp cannot normally give a sufficiently high CMRR for biopotential electrode based measurements, and so it is combined with two buffer op-amps into a construction called an instrumentation amplifier, detailed in the next section.

3.3.3 The Instrumentation Amplifier

There are two major problems with using a single differential amplifier in biomedical measurements. The first, outlined above, is that the CMRR is not high enough to completely eliminate very high common mode signals, which limits the gain that could be applied. The second is that the input impedance of a single-stage differential amplifier is not sufficiently high to buffer it from the effects of a change in the output impedance of the sensor, for example an electrode slowly becoming detached from the body. The most common method of both increasing the CMRR and also providing a high input impedance is to use two additional op-amps at the two inputs to the differential amplifier to form what is termed an instrumentation amplifier. This provides a CMRR of 80 to 90 dB or higher, as well as a very high value of the input impedance, typically 100 to 1000 MΩ.

A schematic of an instrumentation amplifier is shown in Figure 3.25. It consists of a differential amplifier with the inputs provided by two non-inverting amplifiers joined by a common resistor, R_{gain}, which can be used to control the overall gain.

The circuit can be analyzed via superposition. First the voltage source V_2 is set to zero, and V_{out} is represented as $V_{out,1}$, with its value derived from the equation for a simple differential amplifier:

$$V_{out,1} = \frac{R_3}{R_2}(V' - V'') \tag{3.47}$$

Since the non-inverting terminal of the lower op-amp is now at ground, the value of V' can be calculated from the equation for a non-inverting op-amp:

Figure 3.25

(a) Circuit diagram of the
three op-amp
instrumentation amplifier.
(b) Realization in an
integrated circuit design
with voltage inputs and
outputs and connections
for the gain resistor.

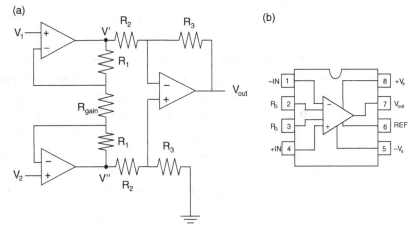

$$V' = V_1 \left(1 + \frac{R_1}{R_G}\right) \tag{3.48}$$

The voltage across R_1 in the lower op-amp is given by $-V''/R_1$, and this must be equal to the voltage across R_1 and R_{gain} (R_G) due to V':

$$V'' = -V' \frac{R_1}{R_1 + R_G} \tag{3.49}$$

Combining these equations, one gets:

$$V_{out,1} = V_1 \frac{R_3}{R_1} \left(1 + \frac{2R_1}{R_G}\right) \tag{3.50}$$

In the second superposition step, the voltage V_1 is set equal to zero. Performing a similar analysis to calculate $V_{out,2}$ gives:

$$V_{out,2} = -V_2 \frac{R_3}{R_1} \left(1 + \frac{2R_1}{R_G}\right) \tag{3.51}$$

Adding $V_{out,1}$ and $V_{out,2}$ together gives the final expression in terms of the differential input $V_2 - V_1$:

$$\boxed{V_{out} = \left(1 + \frac{2R_1}{R_{gain}}\right) \frac{R_3}{R_2} (V_2 - V_1)} \tag{3.52}$$

3.4 Active Filters

By integrating inductive and capacitive elements into the feedback loop or the inputs to the op-amp, the signal can be filtered and amplified in one integrated device termed an active filter. Active filters can be designed with different orders, first-, second-, third- etc., and with different topologies, such as Butterworth and Chebyshev. The most common topology is called a Sallen–Key, which is covered in section 3.4.2.

3.4.1 First-Order Low-Pass, High-Pass and Band-Pass Active Filters

Figure 3.26 shows how a first-order low-pass active filter can be designed using a series RL module at the input to the negative terminal of the op-amp, with a resistor in the feedback loop. At low frequencies the inductor appears as a short circuit and so the circuit reduces to a simple inverting op-amp with gain given by $-R_f/R_{in}$. At high frequencies the inductor appears as an open circuit and V_{out} becomes zero.

Applying the usual condition that there is no current entering the op-amp:

$$\frac{V_{in}}{R_{in} + j\omega L_{in}} + \frac{V_{out}}{R_f} = 0 \tag{3.53}$$

Then the transfer function is given by:

$$\frac{V_{out}(j\omega)}{V_{in}(j\omega)} = -\frac{R_f}{R_{in} + j\omega L_{in}} = -\frac{R_f}{R_{in}(1 + j\omega L_{in}/R_{in})} = -\frac{R_f}{R_{in}} \frac{1}{1 + \frac{j\omega}{(R_{in}/L_{in})}} \tag{3.54}$$

The Bode plots are shown in Figure 3.27. Notice that the effect of the negative sign in the transfer function is to add 180° to the phase plot.

A first-order high-pass active filter can be formed simply by replacing the series RL module by a series RC configuration, as shown in Figure 3.28. At low frequencies C_{in} appears as an open circuit and V_{out} is zero. At high frequencies the capacitor is a short circuit and therefore the gain is given by $-R_f/R_{in}$.

The transfer function is given by:

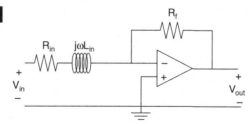

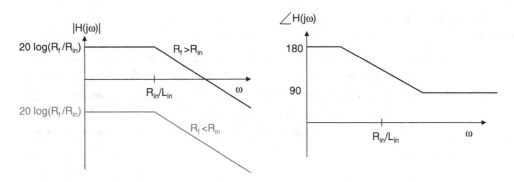

Figure 3.27   Magnitude and phase Bode plots for the active low-pass filter in Figure 3.26.

Figure 3.28

Circuit diagram for an active first-order high-pass filter.

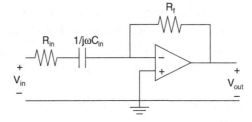

$$\frac{V_{out}(j\omega)}{V_{in}(j\omega)} = -\frac{R_f}{R_{in} + 1/j\omega C_{in}} = -\frac{j\omega C_{in} R_f}{1 + \dfrac{j\omega}{\left(\dfrac{1}{1/R_{in}C_{in}}\right)}} \tag{3.55}$$

The individual components and overall Bode plots are shown in Figure 3.29.

Example
Design an active high-pass filter with a gain of +32 dB in the passband. Using an input resistance R_{in} of 1 kΩ and a critical frequency of 2500 rad/sec, what should the capacitance value be?

Solution
The first step is to convert the gain in dB to a linear scale, which gives a gain factor of 40. Therefore $R_f/R_{in} = 40$ and $R_f = 40$ kΩ. The cut-off frequency is given by $1/(R_f C_{in})$ and so $C_{in} = 0.01$ μF.

In order to design an active band-pass filter design, the circuit must contain two reactive components. Figure 3.30 shows one example containing two capacitors. At low frequencies V_{out} approaches zero due to C_{in} appearing as an open circuit.

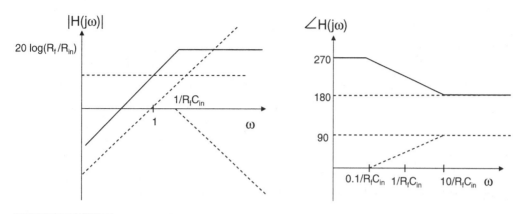

Figure 3.29 Magnitude and phase Bode plots for the active high-pass filter in Figure 3.28. Dotted lines indicate individual components from the expression in equation (3.55), and the solid line represents the total transfer function.

Figure 3.30

Circuit diagram for an active first-order band-pass filter.

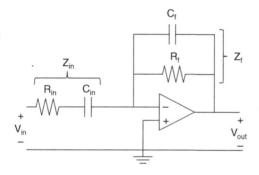

At high frequencies C_f appears as a short circuit and again the output voltage approaches zero.

The transfer function for the band-pass filter in Figure 3.30 is given by:

$$\frac{V_{out}(j\omega)}{V_{in}(j\omega)} = -\frac{Z_f}{Z_{in}} = -\frac{R_f}{1 + j\omega R_f C_f}\frac{1}{R_{in} + \frac{1}{j\omega C_{in}}} = -\frac{j\omega R_f C_{in}}{(1 + j\omega R_{in} C_{in})(1 + j\omega R_f C_f)}$$

$$(3.56)$$

The corresponding individual components and overall transfer function of the Bode plots are shown in Figure 3.31.

The characteristics of single op-amp circuits considered so far can be generalized with respect to their resistive and reactive components in the amplifier inputs and feedback loop, as shown in Figure 3.32 and Table 3.6 (note that the summing and differential amplifiers have geometries that differ from Figure 3.32 and so are not included in Table 3.6).

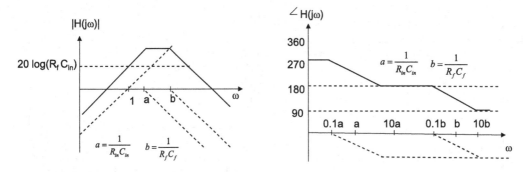

Figure 3.31 (left) Magnitude and (right) phase Bode plots for the band-pass filter shown in Figure 3.30.

Figure 3.32

Generic layout of a single op-amp circuit.

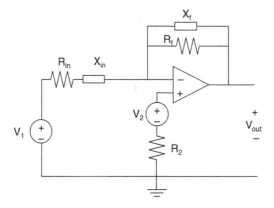

3.4.2 Higher Order Butterworth, Chebyshev and Sallen–Key Active Filters

There are a number of different strategies and geometries that can be used to design higher order active filters. As with passive filters, the most simple approach is to cascade first-order active filters in series, as shown in Figure 3.33. In the circuit shown in Figure 3.33, each individual op-amp circuit represents a first-order inverting, low-pass filter. The overall transfer function is given by the multiplication of the two individual ones.

$$\frac{V_{out}(j\omega)}{V_{in}(j\omega)} = \left[-\frac{2000}{1000} \left(\frac{1}{1 + j\omega(2000)(20x10^{-6})} \right) \right] \left[-\frac{20000}{10000} \left(\frac{1}{1 + j\omega(20000)(1x10^{-6})} \right) \right]$$

$$= \frac{4}{(1 + j\omega/25)(1 + j\omega/50)} \tag{3.57}$$

Table 3.6 Voltage transfer functions for different combinations of resistive and reactive elements in a generic single op-amp circuit shown in Figure 3.32

	R_{in}	X_{in}	R_f	X_f	R_2	V_2	V_1	V_{out}
Inverting	R_{in}	–	R_f	–	–	–	V_{in}	$-\frac{R_f}{R_{in}}V_{in}$
Non-inverting	R_{in}	–	R_f	–	–	V_{in}	–	$\left(1+\frac{R_f}{R_{in}}\right)V_{in}$
Buffer	–	–	–	–	–	–	V_{in}	V_{in}
Differential	R_{in1}	–	R_f	–	R_{in2}	V_2	V_{in}	$\left(1+\frac{R_f}{R_{in}}\right)\left(\frac{R_2}{R_1+R_2}\right)V_1 - \frac{R_f}{R_{in}}V_2$
Integrating	R_{in}	–	–	C_f	–	–	V_{in}	$\frac{1}{R_{in}C_f}\int V_{in}dt$
Differentiating	–	C_{in}	R_f	–	–	–	V_{in}	$-R_f C_{in}\frac{dV_{in}}{dt}$
Low-pass active filter	R_{in}	L_{in}	R_f	–	–	–	V_{in}	$-V_{in}(j\omega)\frac{R_f}{R_{in}\left(1+j\omega R_{in}/R_f\right)}$
High-pass active filter	R_{in}	C_{in}	R_f	–	–	–	V_{in}	$-V_{in}(j\omega)\frac{j\omega C_{in}R_f}{1+j\omega/(R_{in}C_{in})^{-1}}$
Band-pass active filter	R_{in}	C_{in}	R_f	C_f	–	–	–	$-V_{in}(s)\frac{R_f j\omega C_{in}}{\left(1+j\omega R_{in}C_{in}\right)\left(1+j\omega R_f C_f\right)}$

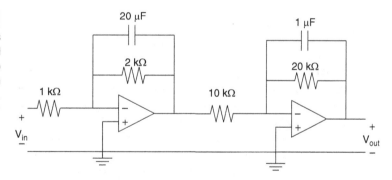

Figure 3.33

Second-order low-pass active filter implemented by cascading together two first-order circuits in series.

Figure 3.34 shows the individual elements of the magnitude and phase Bode plots. Between frequencies of 25 and 50 rad/sec the transfer function decreases at −20 dB/decade, and at frequencies above 50 rad/sec the slope becomes −40 dB per decade.

Cascading active low-order filters to construct higher order filters has the same issues as for passive filters. In practice, the most common topology used for active filters is termed a Sallen–Key circuit, the general form of

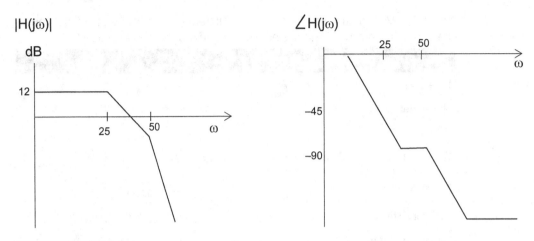

Figure 3.34 Magnitude and phase Bode plots for the cascaded op-amp circuit shown in Figure 3.33.

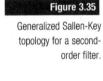

Figure 3.35

Generalized Sallen-Key
topology for a second-
order filter.

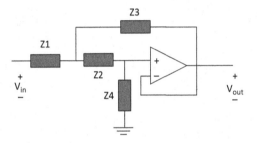

which is shown in Figure 3.35. This type of filter can be designed to have the characteristics of a Butterworth, Chebyshev, Bessel, elliptical or any other filter that might be needed. There are several reasons for its widespread use including:

(i) this configuration shows the least dependence of the overall filter performance on the performance of the particular op-amp. In particular, since it is not based on an integrator design, the gain-bandwidth requirements of the op-amp are minimized

(ii) it can operate with very low gains, which means that the closed-loop dynamics are unconditionally stable

(iii) the ratios of the largest to smallest resistor and the largest to smallest capacitor values are small, which is good for manufacturability

(iv) its use of the minimum number of lumped element components.

The transfer function for the circuit shown in Figure 3.35 is given by:

Figure 3.36

Configurations of low-pass, high-pass and band-pass Sallen–Key active filters.

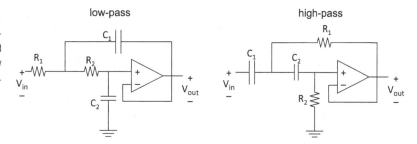

Figure 3.37

Third-order active low-pass filter constructed according to a Butterworth topology.

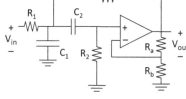

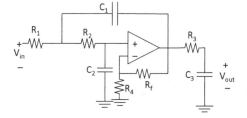

$$\frac{V_{out}(j\omega)}{V_{in}(j\omega)} = \frac{Z_3 Z_4}{Z_1 Z_2 + Z_3(Z_1 + Z_2) + Z_3 Z_4} \tag{3.58}$$

Depending upon the values of Z_1, Z_2, Z_3 and Z_4 different behaviour can be produced, as shown in Figure 3.36.

Higher order filters are commonly synthesized based on second-order filters. For example, Figure 3.37 shows a third-order Butterworth low-pass filter, with a second-order filter placed at the inputs to the op-amp, and a first-order low-pass passive filter after the output.

3.5 Noise in Electrical Circuits

In all considerations for filter and amplifier design up to now, the issue of noise introduced by these components has not been discussed. However, all of the individual components that make up the amplifiers and filters produce some noise. For example, resistors generate a noise voltage given by:

$$V_{R,noise} = \sqrt{4kTR(BW)} \qquad (3.59)$$

Often a device is defined in terms of its noise factor (F), which is the ratio of the SNR at the input divided by the SNR at the output:

$$F = \frac{SNR_{input}}{SNR_{output}} \qquad (3.60)$$

Another commonly used parameter is the noise figure (NF), which is a logarithmic representation of the noise factor:

$$NF(dB) = 10\log F \qquad (3.61)$$

In terms of an amplifier, the output signal is given by the input signal multiplied by the gain. The output noise is given by the input noise (from the sensor) plus the noise of the amplifier, all multiplied by the gain. Therefore the value of F is given by:

$$F = 1 + \frac{Noise_{amplifier}}{Noise_{input}} \qquad (3.62)$$

There are many sources of noise in an amplifier, shown schematically in Figure 3.38.

In general, the signal should be amplified before it is filtered, if this is possible, the reason being that this maximizes the SNR of the signal entering the ADC. In order to understand the reason for this it is useful to analyze the receiver chain in

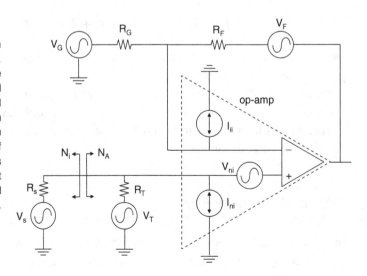

Figure 3.38

Noise sources present in an operational amplifier. Each of the resistive elements Rs, R_G, R_T and R_F have an associated noise voltage, which corresponds to equation (3.59). The op-amp itself can be considered as having a noise current (split into two parts I_{ni} and I_{ii}) and a noise voltage V_{ni}.

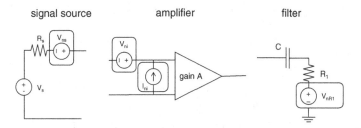

Figure 3.39 Each component in the chain contributes a noise voltage, V_n, and in the case of the amplifier also a noise current (I_n). The noise sources are enclosed in the solid boxes.

terms of the stage-by-stage SNR, which is most easily performed using the concepts of noise factor and noise figure. A noise factor of 1.2, representing a 20% degradation in SNR at the output compared to the input, corresponds to a noise factor of 0.8 dB. As was covered previously, there are several potential sources of noise in electronic circuits, including Johnson thermal, flicker and shot noise. Each component in the detector chain constitutes an extra source of noise, as shown schematically in Figure 3.39 where the noise sources are represented as extra voltage and current sources.

The equivalent noise factor (F_{eq}) of the entire chain (source + amplifier + filter) is given by:

$$F_{eq} = F_1 + \frac{F_2 - 1}{G_1} + \frac{F_3 - 1}{G_1 G_2} + \cdots + \frac{F_n}{G_1 G_2 \ldots G_{n-1}} \quad (3.63)$$

In Figure 3.39 F_1 is the noise figure of the source signal entering the amplifier, F_2 the noise figure and G_1 the gain of the amplifier, and F_3 the noise figure of the filter. Since the amplifier gain is high and the filter has no gain, Equation (3.63) shows that the overall noise figure for the receive chain is minimized if the signal is first amplified and then filtered rather than vice versa. Equation (3.63) also shows that the first stage of the chain is the primary contributor to the overall noise factor, and so the amplifier should be designed to have as low noise voltages and currents as possible.

3.6 Examples of Signal Amplification and Filtering

In this final section, the amplifier and filter design for two of the transducers covered in Chapter 2, namely the pulse oximeter and glucose sensor, are outlined. The starting point for each is the voltage signal produced by the sensor, which may contain significant components of interference signals.

3.6.1   Signal Conditioning in the Pulse Oximeter

As covered in section 2.3 the output of the pulse oximeter is a current that is directly proportional to the amount of light transmitted through the tissue. The components of the device already discussed in Chapter 2 are shown in grey in Figure 3.40. The current i_{out} is on the order of a few milliamps. A block diagram of two different systems containing the remaining components of the device is shown in Figure 3.40.

As stated in section 2.3.3 the current from the photodiodes contains a very large DC component and a much smaller AC component, which is the actual signal, plus noise interference from 50/60 Hz light pollution. A schematic of the signal is shown in Figure 3.41(a). The first step in both configurations in Figure 3.40 converts the current into a voltage using a transimpedance amplifier, which is basically an integrating amplifier with the input resistance set to zero, as shown in Figure 3.41(b).

The transfer function of the transimpedance amplifier is given by:

$$\frac{V_{out}}{I_{PH}} = -\frac{R_F}{1 + j\omega C_F R_F} \tag{3.64}$$

The output of the transimpedance amplifier is ~10 mV in terms of the AC signal of interest on top of a DC term of ~1 volt. The DC term needs to be removed, as does interference from light at 50/60 Hz from the AC term. Figure 3.40(a) shows a simple scheme in which analogue passive and active filters can be used to remove the DC and 50/60 Hz contributions: the design of the filters is shown in Figure 3.42 [5, 6]. The filtered and amplified signal then passes to the ADC, which is covered in section 4.5.1.

In practice, most pulse oximeters use the power of digital filtering and digital signal processing to reduce the effects of interference signals. One scheme is outlined in Figure 3.40(b). The output of the transamplifier needs to be maintained within a strict range, and this is performed via a feedback loop that changes the input currents to the two different LEDs, as shown. In order to remove the DC component, a DC tracking filter extracts the DC component and then feeds it into the final differential amplification stage. The output of this amplifier therefore contains only the desired AC component.

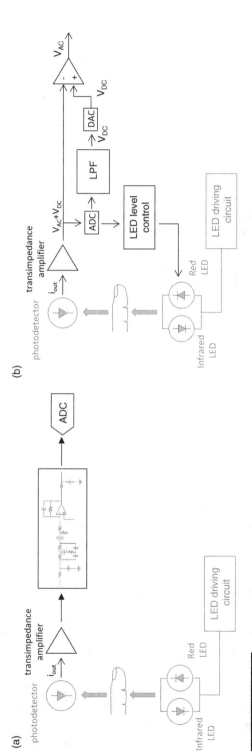

Figure 3.40
Block diagram of a pulse oximeter. The grey components correspond to the sensor already described in Chapter 2, and the black components to the amplification and filter components. (a) Circuit including a transimpedance amplifier followed by passive and active filters. (b) Circuit using digital filtering and active DC tracking. ADC represents an analogue-to-digital converter, and DAC a digital-to-analogue converter.

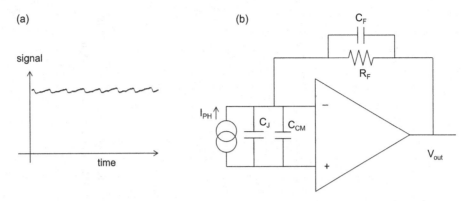

(a) The output signal from the photodiode contains a large DC signal, and a much smaller AC signal from the red and infrared LEDs. (b) A transimpedance amplifier is connected to the output of the photodiode to convert the current I_{PH} to a voltage V_{out}. The feedback capacitance C_F prevents high-frequency oscillations and also forms a low-pass filter with the resistor R_F. The amplifier has an input common mode capacitance C_{CM} and junction capacitance C_J.

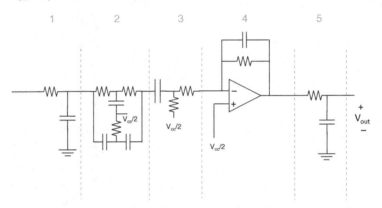

Analogue filtering of the output signal of a pulse oximeter. (1) Low-pass filter with cut-off frequency 6 Hz to remove high-frequency noise; (2) 60 Hz notch filter; (3) high-pass filter with cut-off frequency 0.8 Hz to remove the DC component; (4) first-order active low-pass filter with cut-off frequency 6 Hz and gain of ~30; (5) low-pass filter with cut-off frequency 4.8 Hz.

3.6.2   Amplification and Filtering in a Glucose Sensor

As covered in section 2.4 a glucose sensor produces a DC current of the order of a few microamps, with the magnitude of the current being linearly proportional to the glucose concentration in the blood. Three electrodes are used in the test strip: a reference electrode, a control electrode and a working electrode. Electrons are produced at the working electrode via the mediated chemical reaction.

The first step in processing this current is to convert it into a voltage, using the same type of transimpedance amplifier as used in the pulse oximeter. The reference

(a)

(b)

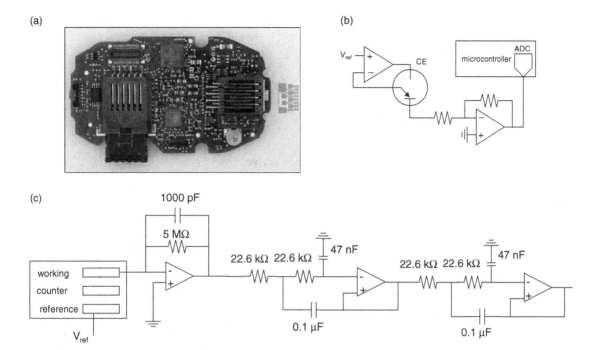

(c)

Figure 3.43 (a) Photograph of a glucose-monitoring strip being placed into the measurement device. The PC board contains the transimpedance amplifier and active filters, as well as an analogue-to-digital converter, which is covered in Chapter 4. (b) The left-hand op-amp fixes the voltage difference between the working electrode and reference electrode, and ensures that no current flows through the reference electrode, while also providing current to the counter electrode, which feeds into the working electrode. (c) Schematic of the current-to-voltage converter and low-pass active Butterworth filter.

electrode is held at a constant voltage (typically –0.4 volts, also called the sensor bias) with respect to the working electrode to drive the chemical reaction, and the counter electrode supplies electrons (current) to the working electrode as shown in Figure 3.43(b). The next step is an active low-pass filter with a cut-off frequency of ~8 Hz to remove high-frequency noise. It is normally a simple second- or fourth-order Butterworth design as shown in Figure 3.43(c). The signal then passes to an ADC, covered briefly in section 4.5.2.

PROBLEMS

3.1 Provide the magnitude and phase Bode plots for the two circuits below:

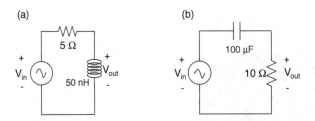

3.2 For the voltage transfer function derived from Figure 3.1(c):

$$H(j\omega) = \frac{j\omega(1/500)}{(1+j\omega/1000)^2}$$

Show in detail each step in deriving the individual components of the magnitude and phase Bode plots shown in Figure 3.6.

3.3 Design a second-order band-pass filter using resistors and capacitors for a microphone for a hearing aid, which allows signals in the range of 1 to 16 kHz to pass through. For the component values that you chose, at what frequency is the transfer function at a maximum?

3.4 Many biomedical instruments pick up noise at 50/60 Hz due to interaction with the main building power supply. Design a four element resonant circuit-based notch filter that will attenuate this signal.

3.5 Determine the frequency-dependent behaviour of the circuit shown below. What type of filter does it represent?

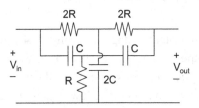

3.6 Design passive Butterworth filters for an EEG system of second, third and fourth orders for an input and output resistance of 50 Ω and a corner frequency of 100 Hz. The topologies can have either shunt or series elements connected on the input side.

3.7 Design a third-order low-pass Chebyshev filter with a 3 dB cut-off frequency of 100 MHz, with a 50 Ω termination.

3.8 In the parallel resonant circuit shown in Figure 3.12(a), reproduced below, the maximum output voltage occurs when the current flowing through the resistor is I_{in}. How much current flows through the inductor, and how much through the capacitor?

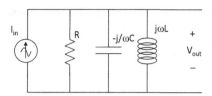

3.9 In practice, many circuits that contain a parallel inductor and capacitor also have an intrinsic resistance associated with the inductor, so the actual circuit can be represented as below. Calculate the input impedance and the resonant frequency of this circuit, and show that for R = 0 the resonance frequency reduces to the expression for a parallel LC circuit.

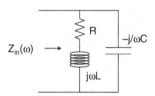

3.10 A circuit consists of two inductors (50 mH and 100 mH), a single resistor (100 Ω) and two capacitors (20 μF and 50 μF). The plots below show magnitude and phase of the input impedance. Construct one possible circuit configuration (there may be more than one solution) that matches the measured data for the different circuits shown in (a) and (b).

(a)

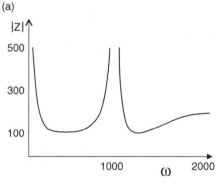

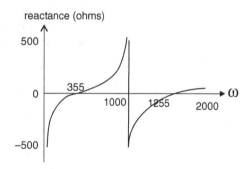

(b)

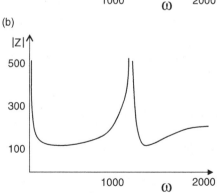

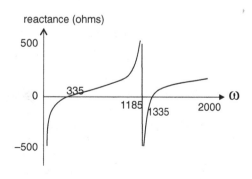

3.11 In the circuit below, the input to the non-inverting input of the op-amp is stepped from 0 to +20 volts in increments of +1 volts. Plot the value of V_{out} as a function of this input voltage.

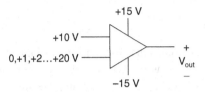

3.12 The circuits for an integrating op-amp, Figure 3.23(a), and an active low-pass filter, Figure 3.28(a), are exactly the same. Explain under which conditions the same circuit can perform these two different tasks.

3.13 Using only resistors and inductors, design an active band-pass filter. Derive the transfer function, $V_{out}(j\omega)/V_{in}(j\omega)$ for your design.

3.14 Derive the expression for the transfer function of a generalized Sallen–Key second-order filter given in equation (3.58).

3.15 Give values for each component of the filter and amplifier circuits used in the pulse oximeter, Figure 3.42. There is no unique solution but components should be chosen with practical values.

3.16 A Butterworth low-pass filter using the smallest possible number of components is to be designed to filter the output of a source at a fundamental frequency of 15 MHz and at harmonics of that frequency. The attenuation at f_0 should be no more than 0.5 dB and the attenuation at the second harmonic, $2f_0$, should be at least 30 dB. Specify the −3 dB cut-off frequency of the filter, and the filter order that will achieve these design goals.

3.17 What are the frequency characteristics of the circuit below?

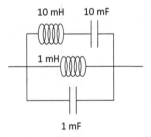

3.18 Show that the following circuit is an exponential amplifier (you need to look up the expression for current through a diode).

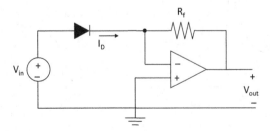

REFERENCES

[1] Kudeki, E. & Munson, D. C. *Analog Signals and Systems*. Prentice Hall; 2008.

[2] Schaumann, R., Xiao, H. & Van Valkenburg, M. *Design of Analog Filters, 2nd edn (Oxford Series in Electrical and Computer Engineering)*. Oxford University Press; 2009.

[3] Clayton, G. B. & Winder, S. *Operational Amplifiers, 5th edn (EDN Series for Design Engineers)*. Newnes; 2003.

[4] Huijsing, J. *Operational Amplifiers: Theory and Design, 3rd edn*. Springer; 2016.

[5] Lee, J., Jung, W., Kang, I., Kim, Y. & Lee, G. Design of filter to reject motion artifact of pulse oximetry. *Comp Stand Inter* 2004; **26**(3):241–9.

[6] Stuban, N. & Niwayama, M. Optimal filter bandwidth for pulse oximetry. *Rev Sci Instrum* 2012; **83**(10):104708.

4 Data Acquisition and Signal Processing

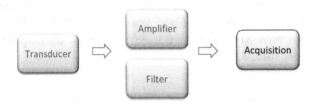

Introduction

After the analogue voltage has been amplified and filtered as described in the previous chapter it is digitized using an ADC. Additional digital filters can be applied to the data before its being transmitted to a display or stored. There are a large number of ADCs available commercially, and the choice of which one to use involves the following considerations [1]:

(i) the frequency bandwidth of the signal
(ii) the required dynamic range and resolution of the ADC
(iii) the maximum voltage that needs to be sampled
(iv) how much power is available for the ADC to operate.

The general principle of digitization is shown in Figure 4.1, in which an analogue signal is converted into an output consisting of digital words in binary format.

The principles of sampling, dynamic range and oversampling are covered in this chapter. The hardware design of ADCs is outlined, together with a brief summary of digital filters required for oversampling. The specifications of different

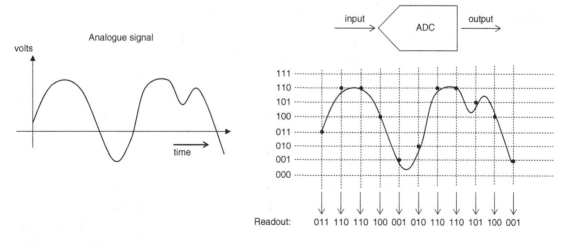

Figure 4.1 In analogue-to-digital conversion a time-domain analogue signal is fed into the ADC, with the output consisting of a digital word. The length (in this case three) of the digital word is determined by the resolution of the ADC.

commercial ADCs and their incorporation into different designs are discussed. Finally, the characteristics of different biomedical signals are outlined, together with commonly used data processing techniques such as cross-correlation, variations of the Fourier transform and filtering to improve the SNR.

4.1 Sampling Theory and Signal Aliasing

The Nyquist–Shannon sampling theory states that in order to correctly sample a signal that consists of frequency components up to a maximum of f_{max}, the sampling rate must be at least $2f_{max}$. For a pure sine wave this means that at least two samples per period must be acquired, e.g. for a signal with components within a bandwidth of DC-20 kHz, the sampling rate should be at least 40 kHz. This corresponds to a time between samples of 25 μs or less. If this criterion is not fulfilled, then aliasing of the signal occurs, as shown in Figure 4.2 [2]. Therefore the particular ADC used in a biomedical device must have a sampling rate that is at least twice the bandwidth of the signal. Ultimately the maximum sampling rate of a particular ADC is determined by the conversion time, i.e. the time it takes to convert an analogue data point into a digital one. This time depends on which type of ADC architecture is used, as covered in section 4.4 of this chapter.

4.2 Dynamic Range, Quantization Noise, Differential and Integrated Non-Linearity

Having determined the required sampling rate of the ADC, the next parameter to consider is the accuracy to which the analogue signal should be approximated by the digitized signal. An ideal ADC would convert a continuous voltage signal into a digital signal with as high a fidelity as possible. The important specifications of an ADC are the voltage range, resolution, maximum sampling frequency and frequency bandwidth, quantization noise and dynamic range.

Voltage range: this is the maximum to minimum voltage range that can be used at the input to the ADC.

Resolution: this specifies the number of different values that the output of the ADC can have, and has units of bits. For an N-bit ADC, the number of different output values is given by 2^N. For example, an 8-bit ADC can produce digital output values from 0 to 255 ($2^8 = 256$), whereas a 16-bit ADC can produce values from 0 to 65 535. The least significant bit (LSB) is defined as one-half the maximum voltage divided by the number of levels. This is sometimes also referred

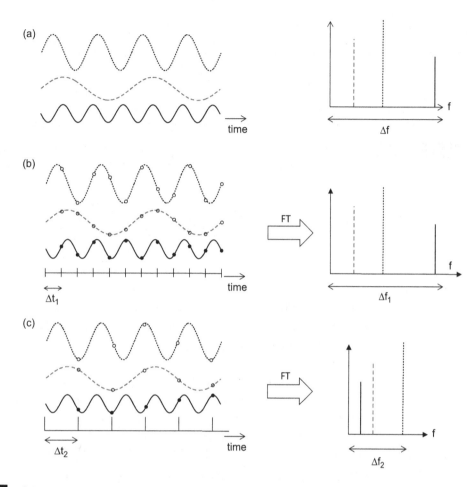

Figure 4.2 Schematic showing the principle of Nyquist sampling and signal aliasing. (a) There are three analogue signals (sine waves), each with a different frequency and amplitude. The total bandwidth of the signals is given by Δf. (b) The signals are sampled according to the Nyquist criterion every Δt_1 seconds. If the digitized signal is Fourier transformed the spectrum shows all three components at the correct frequencies. In (c), the sampling time Δt_2 is twice Δt_1, meaning that the highest frequency signal is sampled less than required by the Nyquist criterion. The highest frequency is now aliased back into the spectrum at an incorrect lower frequency.

to as the voltage resolution of the ADC. The most significant bit (MSB) is defined as one-half of the maximum voltage.

Quantization noise: even ADCs with very high resolution cannot reproduce the analogue signal perfectly. The difference between the analogue input signal and the digitized output is called the quantization error or quantization noise. This error becomes smaller the higher the resolution of the ADC, as shown in Figure 4.3. The values of the quantization error for individual points lie between 0 and $\pm\frac{1}{2}$ of

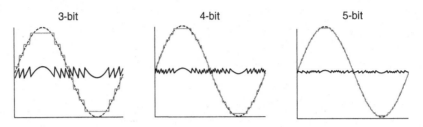

Figure 4.3 The effects of the number of bits on the quantization noise of a digitized signal. The dotted line represents the input analogue signal, the solid grey line the digital output and the solid black line the quantization noise.

the voltage resolution. The root mean square (RMS) quantization noise when sampled at the Nyquist frequency, $e^2_{rms,Nyquist}$, is given by:

$$e^2_{rms,Nyquist} = \frac{\Delta^2}{12} \tag{4.1}$$

where Δ is the voltage resolution. In order to minimize the relative contribution of the quantization errors the input voltage to the ADC should be as large as possible, without 'saturating', i.e. going above the maximum value of the ADC, which leads to signal distortion.

Dynamic range: this represents the ratio of the maximum to minimum input voltages that can be digitized. If the effects of quantization noise are included, then the dynamic range is given by:

$$\text{dynamic range}(dB) = 6.02N + 1.76 \tag{4.2}$$

where N is the number of bits.

Table 4.1 shows the relationship between the resolution, number of levels, least significant bit and dynamic range for a 1 volt ADC.

Table 4.1 Properties of a 1 volt ADC as a function of resolution

Resolution (number of bits)	Number of levels	Least significant bit	Dynamic range (dB)
8	256	3.91 mV	48.2
10	1 024	977 μV	60.2
12	4 096	244 μV	72.3
14	16 384	61 μV	84.3
16	65 536	15.3 μV	96.3
20	1 048 576	954 nV	120.4
24	16 777 216	59.5 nV	144.5

Example

An analogue signal of 526.35 mV is to be digitized using (a) a four-bit, (b) a six-bit, and (c) an eight-bit ADC, each with a voltage range of 1 volt. Calculate the digitized signals (digital word) for the three cases, and the percentage error compared to the actual analogue signal.

Solution

(a) For a four-bit ADC the first bit must be 1 (representing 0.5 volts), which leaves 26.35 mV to be digitized. The second bit is 250 mV, the third 125 mV and the fourth 62.5 mV. In each case, the value 26.35 mV is much smaller than the level, and the error is smaller if a zero is assigned than a one. Therefore the digital word is 1000, with an error of 26.35 mV, which is equivalent to ~5% (100 × 26.35/526.35).

(b) Using a similar argument, the six-bit ADC has levels 500, 250, 125, 62.5, 31.25 and 15.625 mV, which gives a digital word of 100010, with an error of 4.9 mV, which is equivalent to ~1%.

(c) The eight-bit ADC has levels 500, 250, 125, 62.5, 31.25, 15.625, 7.8125 and 3.40625 mV, which gives a digital word of 10000110, with an error of 2.9 mV or ~0.5%.

In practice, the electronic circuits that constitute the ADC introduce non-linear effects, including producing small components at harmonics of the input frequencies, and these are quantified as the total harmonic distortion (THD) of the ADC. The total noise of an ADC is represented by the combination of quantization noise, random thermal noise generated in the electronics of the ADC and the THD. Overall, a quantity referred to as the signal-to-noise-and-distortion ratio (SINAD) is measured for a particular ADC. For a given value of the SINAD the effective number of bits (ENOB) is defined as:

$$ENOB = \frac{SINAD - 1.76}{6.02} \tag{4.3}$$

There are two other measures of error that are typically quoted in ADCs, the differential non-linearity (DNL) and integrated non-linearity (INL), both of which are shown in Figure 4.4. For an ideal ADC, the output is divided into 2 N uniform steps. Any deviation from step uniformity is called DNL and is measured in units of LSBs. For an ideal ADC, the DNL is 0 LSB, but in practice the value should be less than 1 LSB otherwise the performance of the ADC is severely compromised (see Problem 4.4). The INL can be estimated by calculating the cumulative sum of DNL up to that point. Again, the

(a)

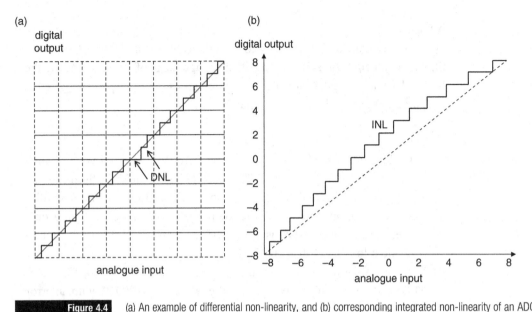

(b)

Figure 4.4 (a) An example of differential non-linearity, and (b) corresponding integrated non-linearity of an ADC.

maximum value of the INL should be less than 1 LSB. The INL is sometimes quoted in units of parts per million (ppm) of the maximum voltage of the ADC.

4.3 Electronic Building Blocks of Analogue-to-Digital Converters

The basic hardware components of an ADC are shown in Figure 4.5. There are a number of different ADC architectures, and associated circuitries, that can be used to digitize the signal. These include successive approximation registers, 1-bit or multi-bit pipeline ADCs and delta-sigma oversampling ADCs. The choice of architecture depends primarily upon the required resolution and sampling rate, as well as the power consumption of the ADC: the latter is particularly important if the device is to be implanted inside the body and run off a battery. In general, there is a trade-off between high resolution and high sampling rates. Before considering each architecture in detail, there are a number of circuit modules that are common to all designs. The first component is a low-pass filter, which removes noise and interfering signals above the maximum frequency of interest: the design of low-pass filters has been described in the previous chapter. The other common modules shown in Figure 4.5 are discussed in the following sections.

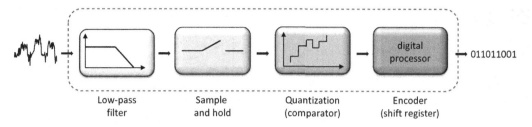

Low-pass   Sample   Quantization   Encoder
filter   and hold   (comparator)   (shift register)

Figure 4.5   Block diagram of an ADC.

4.3.1   Sample-and-Hold Circuits

The objective of the sample-and-hold (S/H) circuit is to sample the analogue signal and hold this sample while the rest of the circuitry in the ADC derives the equivalent digital output. The S/H circuit is designed to have a high accuracy, to function on a timescale of the order of nanoseconds, and have low power consumption and power dissipation. Figure 4.6 shows the principle of the S/H circuit.

The S/H circuit can be designed with or without feedback: feedback gives a higher accuracy but is slower. The two classes of open-loop or closed-loop circuits are shown in Figure 4.7 in their most basic forms. Both circuits are based on buffer op-amps, as covered in Chapter 3. In Figure 4.7(a) switch S_1 is closed during the sampling time and the holding capacitor charges up to the input voltage V_{in}. The switch is often a junction field-effect transistor (JFET), which produces a circuit time-constant τ given by:

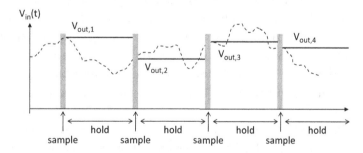

Figure 4.6   Mode of operation of a sample-and-hold circuit. The analogue input signal is shown as a dotted line and sampling times displayed as grey bars. At the first sampling time a value $V_{out,1}$ is recorded and held by the circuit for subsequent processing by the ADC. This value does not change, irrespective of the changing analogue voltage, until the second sampling time, when the value held changes to $V_{out,2}$.

Figure 4.7

(a) Open-loop S/H circuit
with a single switch, S_1.
The switch is closed
during the sampling time
and the holding capacitor
charges up to the voltage
V_{in}. The switch is opened
during the holding time.
(b) A closed-loop S/H
circuit in which the
feedback loop
compensates for any DC
errors in the two
amplifiers.

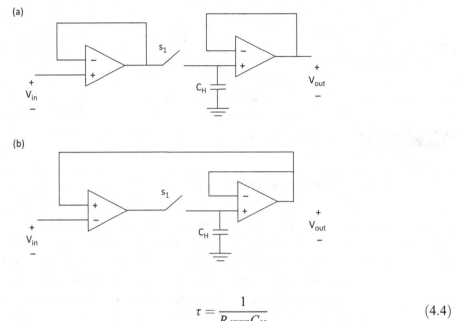

$$\tau = \frac{1}{R_{JFET}C_H} \tag{4.4}$$

where R_{JFET} is the input resistance of the JFET and C_H is the holding capacitance. There is also an additional contribution to the overall time-constant from the slew-rate of the op-amp, which may be of the order of 10 to 50 volts per microsecond. The switch is opened during the holding time, isolating the capacitor, and the voltage V_{in} is maintained across C_H. Buffer op-amps are used for two different reasons. First, when in sample mode the charging time of the hold capacitor depends on the source impedance of the input. A large source impedance would result in a long RC time constant, leading to a long charging time and slow sampling rate. With a buffer op-amp the acquisition time is independent of the source impedance, and is relatively short due to the low output impedance of the op-amp. Second, when in hold mode the hold capacitor discharges through the load at a rate that again depends on the load characteristics. With a high input impedance op-amp the leakage current, which reduces the held voltage over time, is very low.

Figure 4.7(b) shows a closed-loop circuit. The relative advantages of the open-loop and closed-loop configurations relate to speed and accuracy, respectively. In the open-loop circuit the acquisition time and settling time are short because there is no feedback between the buffer amps. However, any DC errors in the two amplifiers add up, so the accuracy is reduced. In the closed-loop configuration any DC errors are compensated, although the speed is somewhat reduced by the feedback loop.

4.3.2 Comparator Circuits

The second module found in every ADC is a comparator circuit, which in its simplest form consists of a single op-amp shown in Figure 4.8(a). The output of the S/H circuit forms the input to the comparator. The input voltage from the S/H circuit is compared with a reference voltage: if the value of the input voltage is higher, then the output voltage is given by $+V_{cc}$, and if the value is lower then $V_{out} = -V_{cc}$, as shown in Figure 4.8(b). Although this simple circuit works in theory, in practice there are several effects that need to be considered. The first is the presence of noise, which means that if V_{in} is very similar to V_{ref}, then any noise on the input voltage will cause the comparator to oscillate rapidly between $+V_{cc}$ and $-V_{cc}$. The second is the fact that op-amps are normally designed with the condition that $V_+ = V_-$ i.e. there is only a very small differential voltage at the inputs. In the case of a comparator there may be quite a large voltage difference $(V_{in} - V_{ref})$ between the two inputs, which can saturate different stages (transistors) within the op-amp, meaning that any switch in the output from $+V_{cc}$ to $-V_{cc}$ will be slow as the individual components of the op-amp require significant time to desaturate, resulting in a low op-amp slew rate. One solution is to add a feedback loop to the amplifier circuit, as shown in Figure 4.8(c), so that the threshold values are now changed to:

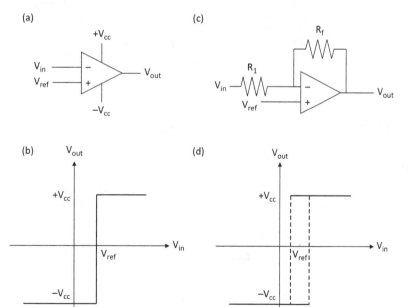

Figure 4.8

(a) Simple op-amp configuration, which compares the input voltage from the S/H circuit with a reference voltage. (b) The output voltage is either $+V_{cc}$ or $-V_{cc}$ depending upon whether V_{in} is greater or less than V_{ref}. (c) An improvement to the design includes a feedback hysteresis loop. (d) The output voltage with the hysteresis values set by the two resistors in equation (4.5).

$$V_{pos,thresh} = V_{ref}\left(\frac{R_1 + R_f}{R_f}\right) + V_{cc}\left(\frac{R_1}{R_f}\right)$$

$$V_{neg,thresh} = V_{ref}\left(\frac{R_1 + R_f}{R_f}\right) - V_{cc}\left(\frac{R_1}{R_f}\right) \tag{4.5}$$

Since there are two different thresholds, this results in a hysteresis loop, as illustrated in Figure 4.8(d).

4.3.3 Shift Register Circuits

The final common hardware component of the ADC is a shift register, a device that produces a discrete delay of a digital signal or waveform. An n-stage shift register delays input data from the comparator by n discrete cycles of an internal clock. For example, a four-stage shift register delays the input data by four clock cycles before moving them to the output data.

In general, data can be fed into or read out of a shift register one block at a time (serial configuration) or many blocks simultaneously (parallel configuration). These operations can be split into four different modes shown in Figure 4.9.

(i) Serial-in-parallel-out (SIPO) – the register is loaded with serial data, one bit at a time, with the stored data being available at the output in parallel form.

(ii) Serial-in-serial-out (SISO) – the data is shifted serially in and out of the register, one bit at a time under clock control. This is the most common form

Figure 4.9

Different types and modes of operation of shift registers.

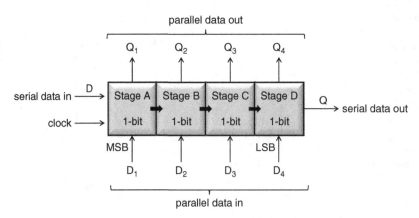

of shift register for ADCs, and acts as a temporary storage, or a time delay, device for the data.

(iii) Parallel-in-serial-out (PISO) – parallel data is loaded into the register simultaneously and is shifted out of the register serially one bit at a time under clock control.

(iv) Parallel-in-parallel-out (PIPO) – parallel data is loaded simultaneously into the register, and all elements are transferred to their respective outputs by the same clock pulse.

Shift registers are used in ADC architectures in which successive comparisons are made between a digitized input voltage (which is held in the S/H circuit) and a series of reference voltages. The shift register records which reference voltage is being compared: this process is covered in more detail in section 4.4.2.

4.4 Analogue-to-Digital Converter Architectures

There are four main types of ADC architectures: flash, successive approximation register (SAR), pipelined and delta-sigma ($\Delta\Sigma$) [1]. In terms of the different architectures, there is a trade-off between the resolution of the ADC and the maximum sampling frequency as illustrated in Figure 4.10(a). For applications requiring very fast sampling with relatively low resolution

(a)

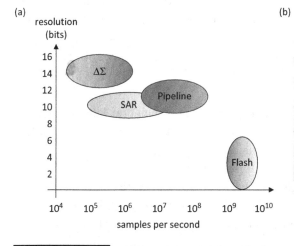

(b)

Architecture	Latency	Speed	Dynamic range
Flash	Low	High	Low
SAR	Low	Low-medium	Medium-high
$\Delta\Sigma$	High	Low	High
Pipeline	High	Medium-high	Medium-high

Figure 4.10 (a) General characteristics of the four main types of ADC in terms of resolution and sampling rates.
(b) Comparisons of the relative performance of these different ADC architectures.

(~8 bits) then the flash architecture is the most appropriate. In practice, flash converters are only used by themselves for very high frequencies above 1 GHz, but they form the basic building block for pipelined ADCs. The most commonly used ADC is the SAR, particularly when multiplexing multiple input channels into a single ADC. This type of architecture can have resolutions of ~18 bits with sampling rates of a few MHz. For applications requiring very high resolution (up to 24 bits) for relatively low frequencies (up to a few hundred hertz) then $\Delta\Sigma$ types are used. There is an overlap between SAR and pipelined ADC performance for frequencies of a few MHz, but above this frequency the pipelined architecture is most commonly used. Figure 4.10(a) only represents typical numbers: there are, for example, a number of higher resolution specialized delta-sigma ADCs on the market. The term latency in Figure 4.10(b) refers to the temporal latency in terms of the combined effect of the number of clock cycles and the clock frequency between the signal entering the S/H circuit and the complete digital word being produced. Latency is important in applications that use multiplexing, i.e. the use of one ADC with many different time-multiplexed inputs.

4.4.1 Flash ADCs

The general principle of flash ADCs is to compare the input analogue voltage at each sampling point with a series of fixed reference values using comparators. The number of fixed reference values is related directly to the resolution of the ADC: an N-bit ADC has 2N-1 comparators and the same number of reference values. As described in section 4.3.2 if the value of V_{in} is less than the reference voltage then the output voltage of the comparator is $-V_{cc}$; if the value is greater than the reference voltage then the output is $+V_{cc}$.

One advantage of a flash ADC is that it has a temporal latency of only one clock cycle between the S/H and the complete digital word. As can be appreciated from Figure 4.11 the major disadvantage is that the number of bits equals the number of comparators, and so a large number of comparators are necessary for high-resolution ADCs. Therefore, as shown in Figure 4.10, a flash ADC is typically used on its own for low-resolution, high-sampling-rate applications, and also as the building block for pipelined ADCs described in section 4.4.3.

Figure 4.11

Basic operation of a two-bit flash ADC with a voltage range of 0 to +4 volts. An input analogue voltage of 1.3 volts gives a comparator output of 001 which enters a decoding network to give the two –bit digital output 01.

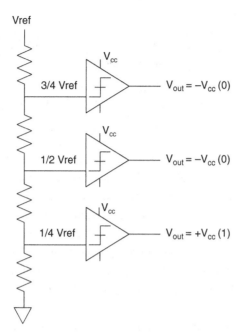

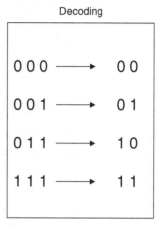

4.4.2 Successive Approximation Register ADCs

Successive approximation register ADCs are based on a design in which the input signal undergoes a series of comparisons with either steadily increasing or steadily decreasing reference values. The input signal is held in the S/H circuit. An N-bit shift register is initially set to its midscale, i.e. the initial value of the reference voltage is one-half the maximum voltage of the ADC. As shown in Figure 4.12 the two signals are compared, and if the input voltage is greater than the reference voltage the most significant bit is set to 1 and the register shifts to the next most significant bit, whereas if the input voltage is less than the reference voltage then the most significant bit is set to zero. The circuit contains a digital-to-analogue converter (DAC), which is essentially the opposite of an ADC, producing an analogue output voltage that is proportional to the input digital value.

Example
Suppose that the input voltage from one sample is 0.37 volts. Assume that the ADC has eight bits with a maximum voltage of 1 volt. Following the flowchart in Figure 4.12, show the steps towards the digital word produced by the ADC.

Solution
The first step is to set the MSB to 1, so the input to the DAC is 10000000. The maximum value of 1 volt corresponds to 11111111, so each of the 256

levels is 3.91 mV. The DAC output of 10000000 corresponds to an output voltage of ~0.5 volts. The output of the comparator shows that $V_{DAC} > V_{in}$, and so the MSB is set to zero. Now the register is shifted to the next MSB, 01000000, which is 0.25 volts. Since now $V_{DAC} < V_{in}$ then the MSB is kept at 1. The next register shift gives 01100000, which is 0.375 volts, and the process continues.

The main advantage of the SAR architecture compared to the flash is that only one comparator is needed, so the design is very simple. In addition, the delay between the digital output and the sampled input is only one clock cycle, i.e. 1/(sampling frequency). The main disadvantage of the SAR ADC is that, as shown by the simplified operation in Figure 4.12, the comparator, DAC and the entire SAR logic must operate at N-times the sampling frequency, where N is the number of bits. Therefore the highest possible sampling frequency of the ADC is only a small fraction of the capabilities of most of its components. This is the fundamental reason why SAR ADCs generally operate with medium sampling rates. The very low latency compared with pipelined and delta-sigma ADCs means that the SAR architecture is very often used in cases where multiple input channels are multiplexed (switched) into a single ADC.

Figure 4.12

Block diagram of an SAR ADC with a single comparator. For each sample of the analogue input signal the value is compared to a decreasing voltage until the input value is greater than the reference value.

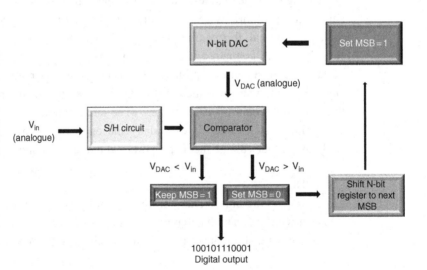

4.4.3 Pipelined ADCs

The concept of a pipelined ADC is to cascade several low-resolution stages together to obtain a high-resolution device [3]. A pipelined ADC architecture has a high resolution compared to flash or SAR designs at the expense of an increased temporal latency. The principle is that the analogue signal passes successively through a series of 'pipelines', each of which digitally encodes a subset of the final digital output signal. The pipelined approach can achieve very high sampling rates without needing a very high clock frequency. Figure 4.13 shows a block diagram of a pipelined ADC.

In order to analyze how the pipelined ADC works, for simplicity consider just the first two stages in an N-bit pipeline. In the first clock cycle the ADC quantizes the first N/2 MSBs. During the second clock cycle, these N/2 MSBs are subtracted from the input signal, and the residual signal is amplified by a factor A to increase the dynamic range of the second stage, and the remaining N/2 bits are digitized. A shift register corrects for the fact that the digital signals from each stage are effectively acquired at slightly different times.

For an N-bit pipelined ADC, the number of comparators needed is $2^{N/2+1}$, which is lower than the 2^N required for a flash ADC. Although many clock cycles are needed to produce the full digitized output, a partial output is

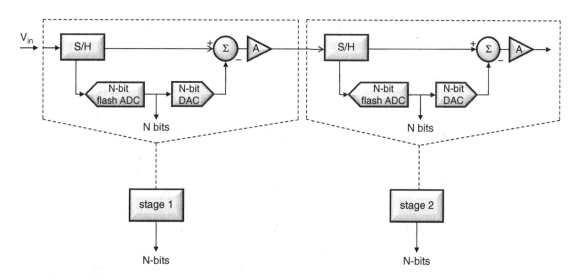

Figure 4.13 Block diagram of a pipelined ADC. Each stage has a digital output of N-bits, with an amplification step of gain A between each stage. The digital resolution is given by the number of stages multiplied by the number of bits per stage.

available every clock cycle. Therefore the overall speed of the pipelined ADC is only limited by the latency through a single stage, which is much lower than for a SAR ADC where the temporal latency corresponds to digitizing the entire analogue input signal. One manufacturing advantage of the pipelined architecture is that the precision requirements for each pipeline stage decrease along the pipeline, i.e. as the stage number increases. So, for the first stage the S/H must be accurate to N-bits, as must the DAC. However, in the second stage, due to the stage-gain A, the second S/H only requires $(N/2 + 1)$ bit accuracy and so on along the pipeline.

4.4.4   Oversampling ADCs

There are a number of issues associated with the Nyquist sampling schemes that are used with the ADC designs outlined in the previous sections. First, the analogue low-pass filter shown in Figure 4.5 cannot be made absolutely 'sharp', and has other problems such as group delays and frequency-dependent phase shifts. This means that either a much larger bandwidth is selected than is theoretically needed (which introduces more noise) or that some degree of amplitude/phase distortion of the signal has to be accepted. Second, very high resolution (>16 bit) ADCs are difficult to produce in SAR and pipelined architectures. Fortunately, there is a method, termed oversampling, which can increase the effective resolution of an ADC. Oversampling is used in the fourth class of ADCs considered here, namely the $\Delta\Sigma$ architecture [4].

4.4.4.1   Principle of Oversampling

Oversampling, as the name suggests, involves sampling the signal at a much higher rate than required by the Nyquist theorem, as illustrated in Figure 4.14(a) and (b). Since the bandwidth of the digitized voltage is much larger than the actual bandwidth of the signals of interest, the lack of sharpness of the low-pass filter is no longer a problem as illustrated in Figure 4.14(b): the associated issues of group delays and phase shifts also effectively disappear.

After the data has been digitized, two additional signal processing steps are performed in software. First, a *digital filter* is applied to the sampled data to select only the bandwidth of interest. Such a digital filter can be designed to have an extremely sharp transition (see section 4.4.4.3). At this stage, the filtered data still has M-times as many data points (where M is the oversampling factor) as would have been acquired without oversampling. The final signal processing step is to 'decimate' the data, a process in which M consecutive data points are

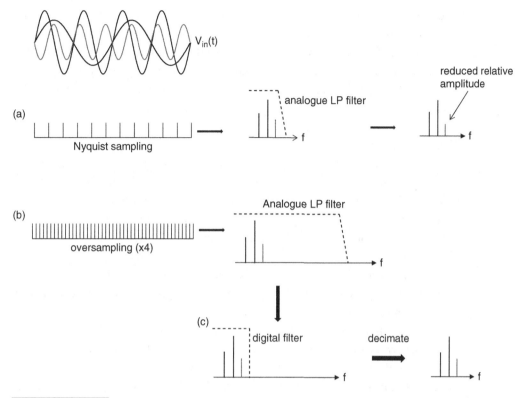

Figure 4.14 Schematic showing the principle of oversampling. The input signal has three different frequencies with three different amplitudes. (a) Sampling at the Nyquist frequency with an analogue low-pass filter can result in attenuation of peaks close to the cut-off frequency of the low-pass filter.
(b) If oversampling is performed then the cut-off frequency of the analogue filter is much higher than that of the signals and so no distortion occurs. (c) After digitization, a digital filter is applied to select the frequency band of interest, and in the final step the signal is decimated, which reduces the contribution of quantization noise.

averaged together to give a single point, thus reducing the data size correspondingly. Since the quantization noise in consecutive data points is random, the overall quantization noise is reduced, with respect to that of a Nyquist sampled signal, by a factor of $\sqrt{M}$ to a value given by:

$$e^2_{rms,oversampled} = \frac{\Delta^2}{12\sqrt{M}} \qquad (4.6)$$

where Δ is the voltage resolution. For every factor of four in oversampling, the equivalent resolution of the ADC increases by 1-bit. Of course, the disadvantage of the oversampling approach is that the ADC must operate at a higher sampling rate, and the digital signal processing unit must also perform digital filtering and decimation of the data. If signal processing needs to be performed in real time, as is the case for example in digital hearing aids, covered in Chapter 7, then the degree of signal processing must be tailored to the acceptable time difference between the input signals being received and the output signals being fully processed.

4.4.4.2 Delta-Sigma ADCs

The majority of oversampling ADCs are of the delta-sigma ($\Delta\Sigma$) configuration. The name comes from the fact that a differential (Δ) signal is fed into a summer (Σ). $\Delta\Sigma$ ADCs can be used for applications requiring high resolution (16 bits to 24 bits) and effective sampling rates up to a few hundred kHz. A block diagram of a first-order $\Delta\Sigma$ modulator that incorporates a negative feedback loop is shown in Figure 4.15.

The sampling rate of the $\Delta\Sigma$ ADC is dictated by the clock frequency, and so can be very high. One of the unique features of the $\Delta\Sigma$ architecture is that it contains an integrator, and this affects the frequency distribution of the quantization noise. In a conventional Nyquist-rate sampling scheme, the quantization noise, shown in Figure 4.16(a), is given by equation (4.3). If an oversampling ADC is used, then the quantization noise is reduced by a factor of the square-root of the oversampling factor, equation (4.6), but the frequency range over which the quantization noise is spread is increased by this oversampling factor. A digital filter (see next section) can be used to remove this noise, as shown in Figure 4.16(b). In a $\Delta\Sigma$ ADC, the frequency characteristics of the quantization noise are different, as shown in Figure 4.16(c), with the noise power increasing

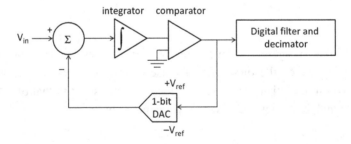

Figure 4.15 Block diagram of a first-order $\Delta\Sigma$ modulator. The output of the modulator is a stream of ones and zeros, i.e. a one-bit data stream. The feedback loop means that the input to the summer from the 1-bit DAC must on average be equal to V_{in}. If V_{in} is zero, i.e. at the mid-scale of the ADC, then there are an equal number of ones and zeros output from the modulator. If V_{in} is positive then there are more ones than zeros, and if V_{in} is negative there are more zeros than ones. The fraction of ones in the output is proportional to V_{in}.

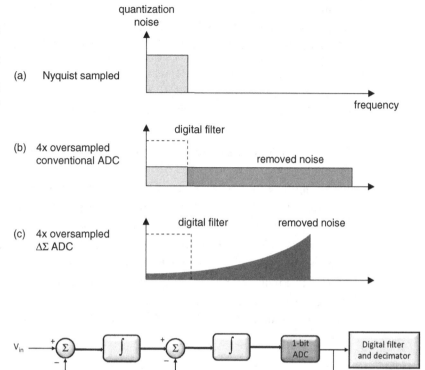

Figure 4.16

The frequency spectrum of quantization noise for: (a) a conventional Nyquist sampling ADC, (b) a conventional oversampling ADC, and (c) a ΔΣ oversampling ADC.

Figure 4.17

Schematic of a second-order ΔΣ modulator.

as a function of frequency. This means that a low-pass digital filter will remove a greater proportion of the noise than if no noise-shaping were present.

In practice, a higher order ΔΣ modulator is normally used. A second-order modulator has better performance in terms of noise shaping. Figure 4.17 shows a 1-bit, second-order modulator that has two integrators instead of one. Increasing the number of integrators in the modulator provides more noise shaping, but greater than third order is difficult to implement. Instead, the 1-bit ADC (the comparator) can be replaced with an N-bit flash converter, and the 1-bit DAC (switch) by an N-bit DAC.

4.4.4.3   Digital Filters

After the signal has passed through the ΔΣ ADC the digitized data passes through a low-pass digital filter [5]. The ideal filter response has a very sharp transition in terms of its magnitude response, with a linear phase across the entire range of frequencies that passes through the filter, meaning that phase distortion and the related group delay are avoided. The most common forms of digital filters are called finite input response (FIR) filters, which produce an output by computing the dot product of a constant coefficient vector with a vector of the most recent input signal values.

$$y[n] = \sum_{k=0}^{N-1} h[k]x[n-k] \tag{4.7}$$

where y represents the output signal, x the input signal and h contains the filter coefficients.

As can be seen from equation (4.7) the required computations make heavy use of what are termed multiply–accumulate operations, and so digital processing chips are especially designed to be efficient with respect to this type of computation. The filter can be designed with any number of 'coefficients', the greater the number the more ideal the behaviour of the filter, as shown in Figure 4.18. However, when the digital filter has to be applied in real time to streaming data (for example with a digital hearing aid), then there is a trade-off between the desired filter characteristics and the number

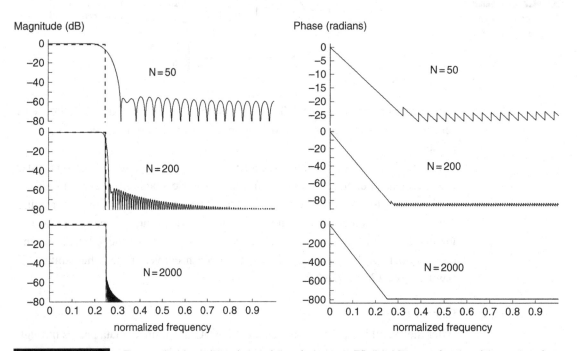

Figure 4.18   The magnitude and phase characteristics of a low-pass FIR digital filter as a function of the number of coefficients used in the design. The normalized cut-off frequency is set to 0.25, corresponding to an oversampling factor of four.

of coefficients defining the filter, with a higher number of coefficients requiring more processing time, resulting in a longer delay in the output signal with respect to the input signal.

The final step is decimation, which as mentioned earlier essentially corresponds to an averaging of M successive data points to produce a single data point, where again M is the oversampling factor.

4.5 Commercial ADC Specifications

The choice of the appropriate ADC, or multiple ADCs, for each medical device is made based primarily upon the required resolution and sampling rate, and the amount of power needed. An ADC from a commercial manufacturer comes with a full and detailed specification sheet, which in addition to the parameters covered in this chapter also includes measures such as spurious-free dynamic range (SFDR), two-tone SFDR and RMS jitter time, which are covered in more specialized literature. The description of ADCs finishes with consideration of appropriate choices for a pulse oximeter and glucose monitor.

4.5.1 ADC for a Pulse Oximeter

Continuing from section 3.6.1, the amplified and filtered analogue voltage V_{AC} (Figure 3.40) enters the ADC. V_{AC} actually contains four separate interleaved measurement data streams: the voltage from the infrared sensor, the ambient voltage of the infrared light without illumination, the voltage from the red-light sensor and the ambient voltage of the red-light without illumination. The sampling time is approximately 250 μs, and therefore a sampling frequency of a few kHz is more than sufficient to fulfil the Nyquist criterion. Given this low sampling frequency, and the small changes in light intensity that must be detected, a high-resolution (usually greater than 18 bits) $\Delta\Sigma$ ADC is often chosen, with an oversampling frequency between 40 and 80 kHz. The serial output of the ADC is then digitally split into four different sections corresponding to the four individual measurements, and then the calculations outlined in section 2.3.2 are performed to produce the SpO_2 values, as well as other measures such as heart rate. The results of these calculations are then displayed on the LCD screen.

4.5.2 ADC for a Glucose Meter

The output voltage of the transimpedance amplifier and low-pass filter circuitry, covered in section 3.6.2, results in a voltage that can be sampled at a relatively low rate and does not have to have extremely high resolution. Given these requirements the most commonly used ADC is a SAR architecture in the range of 12 to 16 bits.

4.6 Characteristics of Biomedical Signals and Post-Acquisition Signal Processing

After the particular signal has been digitized it can be stored and displayed. Since the data is digital, it is possible to perform different signal processing techniques to try to improve the diagnostic quality of the data. This section describes a few of the more common techniques used to process biomedical signals. The particular technique used is highly dependent upon the characteristics of the signal, and so the first section considers the different classes of biosignals that are commonly recorded [2].

4.6.1 Deterministic and Stochastic Signals

Figure 4.19 shows examples of five different biosignals covered in this book [2]. The signals have very different frequency contents and amplitudes, which determine the degree of amplification and filtering required, as seen in previous chapters.

In terms of characterizing the nature of different biosignals the primary classification is whether they are either **deterministic** or **stochastic**. Deterministic signals are those for which 'future' signals can be predicted based upon 'past' signals. For example, ECG signals over a short time-frame are considered deterministic, in that neither the heart rate nor the shape of the ECG trace changes abruptly. In contrast, stochastic signals do not have this property: in Figure 4.19 the EEG and speech signals are clearly stochastic. However, for stochastic signals, even though future signals cannot be predicted based on previous signals, certain properties of the signals are time independent. An example might be an EEG trace, in which the SNR is very low, and therefore the exact future signal cannot be predicted; however, the mean and the standard deviation of the signal remain the same over time.

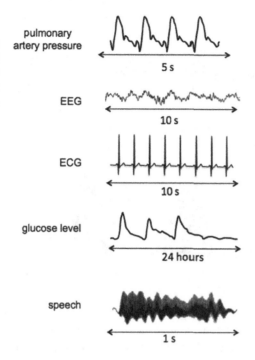

pulmonary
artery pressure

5 s

EEG

10 s

ECG

10 s

glucose level

24 hours

speech

1 s

Biosignals are of course never completely deterministic. There is always some random noise component or physiological variation that renders them non-deterministic. It is, however, very convenient to characterize signals such as the ECG by means of a deterministic function. Deterministic signals, s(t), can be described in terms of their periodicity:

$$s(t) = s(t + nT) \qquad (4.8)$$

where n is an integer number and T is the period of the signal. Blood pressure and the ECG signals are examples of biomedical signals that are the most periodic.

Stochastic signals can be characterized by statistical measures such as their mean and standard deviation over a certain time period. Since stochastic signals cannot be modelled, they can be represented as the probability of a signal occurring in terms of the signals recorded in the time period before. The **probability density function** (PDF) of the signal is a histogram of all of the signal amplitudes within a certain time period. Figure 4.20 shows three PDFs calculated from consecutive time periods of an EEG signal. Stochastic signals can be subdivided into either **stationary** or **non-stationary** signals. A stationary signal is defined as one in which the PDF does not change over time, whereas for a non-stationary signal the PDF can be different over

Figure 4.20

A section of a continuously acquired EEG signal. Three sections of equal time duration are defined, and the PDF of each section calculated. The similarity in the PDFs shows that these particular EEG signals can be considered as stochastic and stationary.

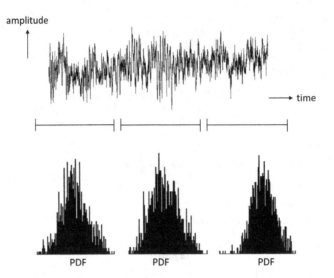

different measurement periods. The data presented in Figure 4.20 shows that this particular EEG signal is stationary.

Finally, stationary stochastic signals can be further subdivided into two classes: **ergodic** and **non-ergodic**. For ergodic signals, the PDF is the same whether measured across multiple signals or within a single signal, which makes it possible to calculate the PDF from a single time-domain signal, rather than having to acquire many different time-domain signals. As an example, an EEG acquired while the patient is asleep is technically a non-stationary signal, i.e. the characteristics change over time as the state of sleep changes. However, if one considers the signal over a short period of time, then it can be treated as stationary. Very often it is also considered to be ergodic, and therefore the processing of single time periods can be used to estimate the PDF and other statistical measures.

Figure 4.21 gives an overview of the most commonly used classifications of biomedical signals.

The characteristics of the **noise** in a particular biomedical instrument depends both upon the physiological basis of the measurement and also on the electronic system used to acquire the signals. Very often interfering signals occur at frequencies different to those of the signals of interest and therefore can be filtered out. However, there is also a component of noise that is associated with the electronics used to make the measurements, i.e. the transducer, the filter, the amplifier and the ADC each introduce a certain amount of noise. A common assumption is that the overall noise level of the measurement system has the characteristics of 'white noise' or 'Gaussian

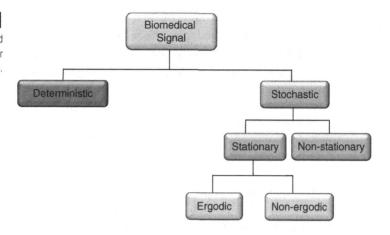

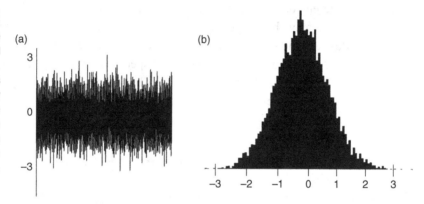

noise', which means that the noise level is independent of frequency. An example is shown in Figure 4.22, which shows that random noise has a Gaussian probability density function. As covered in previous chapters there are also certain devices that have components of noise which are frequency dependent: one such example is flicker noise. However, for most biomedical measurement systems the assumption of Gaussian noise is a good approximation.

4.6.2 The Fourier Transform

Although data is recorded in the time domain, the information content of the digitized signal can also be visualized in the frequency domain. This simplifies the determination of the frequency bandwidth in which the desired signal occurs, and also frequencies that correspond to interfering signals and need to be filtered out. The Fourier transform, F{}, of a time-domain signal s(t) to the corresponding frequency-domain signal S(f) is given by:

$$S(f) \;=\; F\{s(t)\} = \int_{-\infty}^{\infty} s(t)\, e^{-j2\pi ft} dt \tag{4.9}$$

Fourier analysis assumes that the signal is stationary but many biomedical signals do not strictly fit this criterion. However, if the signal is split into short segments then each segment by itself can be considered as a windowed version of a stationary process. These segments can be equal in duration, or different, depending upon the characteristics of the signal. For example, an EEG signal may show a sudden burst of activity during a quiet background, and in this case the signal should be separated into time periods of different length, as shown in Figure 4.23. In contrast in speech processing equal-duration windows of the order of 10 to 20 ms are usually chosen for frequency analyis. For such a windowed time period the short-time Fourier transform (STFT) of the signal is defined as:

$$STFT(\omega, \tau) = F\{s(t)w(t-\tau)\} = \int_{-\infty}^{\infty} s(t)w(t-\tau)e^{-j\omega t}dt \tag{4.10}$$

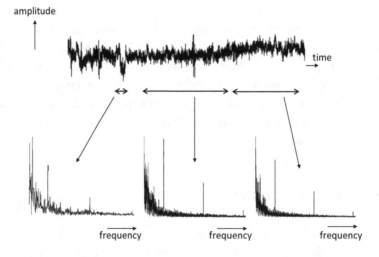

Figure 4.23

An EEG signal that is split into time periods of different length for analysis using the short-time Fourier transform.

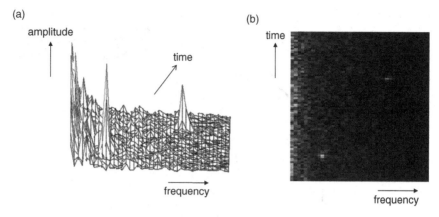

with the FT being performed on a segment that lies in the time period $\tau-(T/2)$ $\leq t \leq \tau+(T/2)$. As an example, Figure 4.23 shows a noisy EEG signal, and its corresponding Fourier transform.

For any given biosignal, the frequency components may change as a function of time, and this can be displayed as a spectral array or density spectral array as shown in Figure 4.24. These types of display are often standard features on systems such as EEG monitors.

4.6.3 Cross-Correlation

Cross-correlation techniques are used to measure the 'similarity' of two signals that might be shifted in time or have the same time-dependent behaviour but occur in different parts of the body. Mathematically, the cross-correlation function (denoted by **) of two functions f(t) and g(t) is given by:

$$f^{**}g = \int\limits_{-\infty}^{\infty} f(t)g(t+\tau)dt \qquad (4.11)$$

An example of the use of cross-correlation in a biomedical device is the measurement of cardiac output using a Swan–Ganz catheter [6]. Cardiac output is defined as the average blood flow through the cardiac muscle over one minute. This invasive measurement is performed using pulmonary artery catheterization (PAC) [7]. A catheter with two thermal sensors is inserted into the pulmonary artery as shown in Figure 4.25(a). Cold sodium

(a) (b)

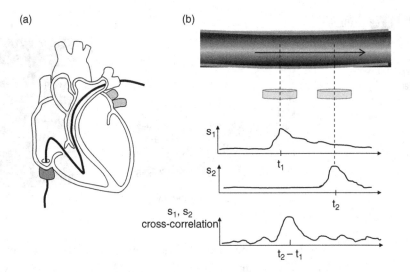

(a) Schematic of a Swan–Ganz catheter, which is used for pulmonary artery catheterization thermodilution measurements of cardiac output. (b) A cold solution of NaCl is injected through the catheter and two temperature measurements are made with thermistors placed along the catheter. By performing a cross-correlation between the two measurements, the time ($t_2 - t_1$) for blood to flow between the two points can be calculated.

chloride solution is injected into the right atrium of the heart via a small duct inside the catheter, and the temperatures are recorded by two sensors placed along the catheter as a function of time. If the signals are ergodic and stationary, and the form of the signal is essentially the same at the two measurement points, then the cross-correlation function of the two signals has its maximum value at a time τ given by:

$$\tau = \frac{L}{v_{av}} \tag{4.12}$$

where v_{av} is the average blood flow velocity and L is the distance between the measurement points. The volumetric flow (Q) can then be calculated if the cross-sectional area (A) of the vessel is known:

$$Q = v_{av}A \tag{4.13}$$

There are many other examples of signal processing involving cross-correlation, including methods of detecting wind noise in hearing aids, which are covered later in Chapter 7.

4.6.4 Methods of Dealing with Low Signal-to-Noise Data

Some biosignals have an intrinsically low SNR: one example is individual measurements of glucose concentration obtained with an implanted biosensor, covered in detail in Chapter 8. In such cases, it is important to consider methods of increasing the SNR, either in data acquisition or in post-acquisition signal processing. This section considers a few very basic methods.

4.6.4.1 Signal Averaging

If the signal is deterministic, and the noise is Gaussian, then a simple method for increasing the SNR is to add successive signals together. The measured signal, S', can be represented as the true signal (S) plus a noise component (N):

$$S' = S + N \tag{4.14}$$

If the noise is Gaussian, as shown in Figure 4.22, with mean value of zero and a standard deviation σ_N, then the SNR for a single measurement is given by:

$$\mathrm{SNR}_{S'} = \frac{|S|}{\sigma_N} \tag{4.15}$$

If K measurements are acquired, added together, and then averaged the total signal S'' is given by:

$$S'' = S + \frac{1}{N} \sum_{k=1}^{K} N_k \tag{4.16}$$

Assuming that the noise for each measurement is uncorrelated, then the SNR for the averaged scans is given by:

$$\mathrm{SNR}_{S''} = \frac{|S|}{\sqrt{\mathrm{var}\left\{ \frac{1}{K} \sum_{k=1}^{K} S + N_k \right\}}} = \sqrt{K} \frac{|S|}{\sigma_N} = \mathrm{SNR}_{S'} \sqrt{K} \tag{4.17}$$

showing that the SNR increases as the square root of the number of averaged measurements. Figure 4.26 shows an example of the effect on the SNR of increased signal averaging.

(a)   (b)   (c)   (d)

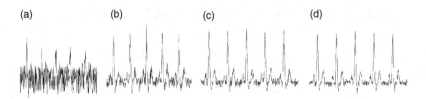

Figure 4.26   Synthetic plot of a noisy ECG trace from a single lead electrode. (a) Single signal, (b) 20 averaged signals, (c) 50 averaged signals, and (d) 100 averaged signals.

4.6.4.2   Moving Average

In cases where individual measurements are noisy and the aim is to determine a general trend in the signal level, then a moving average filter can be used. One of the biomedical devices that uses a moving average is an implanted glucose meter, covered in Chapter 8. The device monitors the glucose level every minute, and can be used to predict when and how much insulin the patient should inject. Individual measurements are noisy, and so if predictions are made from these individual measurements then the overall trend can be obscured: therefore a moving average is used.

If the original data set has values a_1, a_2, a_3 ... a_N, then the n-point moving average s_1, s_2 ... s_{N-n+1} is defined as:

$$s_i = \frac{1}{n} \sum_{j=1}^{i+n+1} a_j \tag{4.18}$$

An example of calculating the moving average of a noisy signal is shown in Figure 4.27. The higher the number of points used for the moving average the higher the overall SNR, but the lower the effective temporal resolution of the measurements.

4.6.4.3   Post-Acquisition Digital Filtering

As described previously, digital filters can be applied on the ADC chip in real time to streaming data. In addition, after the signals have been digitized, other types of filtering can be applied if necessary. For example, some patients exhibit a higher degree of low-frequency motion-induced artefacts than others, and this requires additional post-acquisition filtering to remove these artefacts. In another commonly encountered situation interference from nearby machines is different in different environments, particularly if the measurement device is portable. Very often, additional notch filtering of the 50/60 Hz interference signal may be necessary if the equipment is close to building power lines in the wall or ceiling.

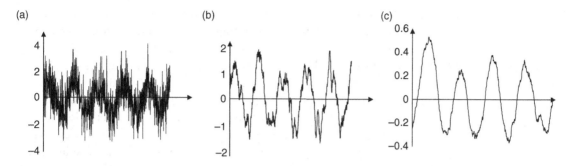

Figure 4.27 Example of calculating a point moving average of a signal with relatively low SNR. (a) Original signal, (b) 15-point moving average, and (c) 150-point moving average.

PROBLEMS

4.1 The sampling rate for a bioinstrument is set at 100 Hz. However, the input signal actually contains frequency components at 25 Hz, 70 Hz, 160 Hz and 510 Hz. Show the frequency spectrum that is output by the system.

4.2 Explain the effects of a signal that has a magnitude higher than the upper limit of an ADC. What does the time-domain signal look like, and what effect does this have on the frequency-domain signal?

4.3 What is the minimum voltage difference that can be measured by a 5 volt, 12-bit ADC?

4.4 Construct an analogue sine-wave input to a 3-bit ADC. Show the output for the DNL characteristics given below.

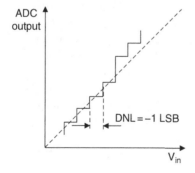

4.5 Sketch the block diagram for a three-bit flash ADC. If V_{ref} is 1 volt, and the input voltage is 0.42 volts, show the output from each comparator in terms of a 0 or 1, and the digital word that forms the output of the circuit.

4.6 Consider the circuit below, which is used for measuring the oscillatory signal in an automatic blood pressure monitor covered in Chapter 2. There are two input signals, the cuff pressure at 0.04 Hz and the oscillation signal at 1 Hz. The maximum input signal of the oscillation signal is 36 mV. The signal is detected using a 5-volt ADC converter. What should the gain of the amplifier be in order to use the full resolution of the ADC?

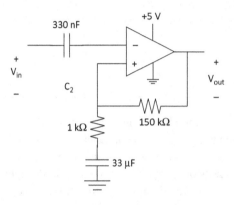

4.7 For the comparator circuit shown in Figure 4.8 derive the positive and negative threshold values for the output voltages in equation (4.5).

4.8 What is the output digital word for a 2.25 volt signal that is fed into a 5 volt 8-bit flash ADC?

4.9 In an 8-bit SAR ADC, with a maximum voltage of 5 volts, the input voltage from the first sample is +1.38 volts. Following the flowchart in Figure 4.12, show the steps towards the digital word output from the ADC.

4.10 By searching the literature describe one type of circuit that can be used to decode the comparator output of a flash ADC to a digital word.

4.11 In the case of a band-limited high-frequency signal, as shown below, the signal can be undersampled, i.e. the sampling can be at a rate less than the Nyquist sampling rate. Assuming in the example below that the sampling rate is set at 1000 Hz, show the resulting signal, explain the process involved and determine what features the filter before the ADC should have.

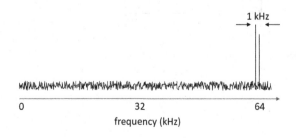

4.12 Considering equation (4.6) explain why for every factor of four in oversampling, the equivalent resolution of the ADC increases by 1-bit.

4.13 Explain the process of noise shaping, i.e. why in a delta-sigma ADC does the noise increase as a function of frequency?

4.14 Based on your knowledge of op-amp circuits, devise a simple circuit for a delta-sigma ADC.

4.15 Investigate an example of using cross-correlation techniques in biomedical signal processing. Explain the sensors used, the measurements performed and the results of the analysis.

4.16 If the noise in a signal is coherent, what is the effect of signal averaging?

4.17 Suppose that the noise shown in Figure 4.22 is displayed in magnitude mode. Plot the probability density function, and calculate the mean and standard deviation of the distribution.

REFERENCES

[1] Pelgrom, M. J. *Analog-to-Digital Conversion*. Springer; 2013.

[2] Semmlow, J. *Signals and Systems for Bioengineers: A MATLAB-Based Introduction*. Academic Press; 2011.

[3] Ahmed, I. *Pipelined ADC Design and Enhancement Techniques (Analog Circuits and Signal Processing)*. Springer; 2010.

[4] Pandita, B. *Oversampling A/D Converters With Improved Signal Transfer Functions (Analog Circuits and Signal Processing)*. Springer; 2011.

[5] Allred, R. *Digital Filters for Everyone*. Creative Arts & Sciences House; 2015.

[6] Yelderman, M. Continuous measurement of cardiac output with the use of stochastic system identification techniques. *J Clin Monit* 1990; **6**(4):322–32.

[7] Chatterjee, K. The Swan–Ganz catheters: past, present, and future: a viewpoint. *Circulation* 2009; **119**(1):147–52.

5 Electrocardiography

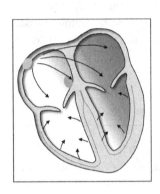
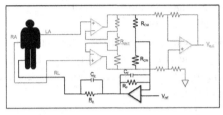
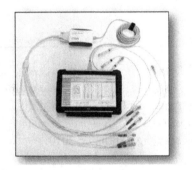
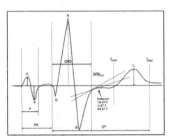

Introduction

An ECG, also sometimes referred to as an EKG from the original German word 'electrokardiogram', measures the electrical activity of the heart [1]. This electrical activity produces the contractions and relaxations of the cardiac muscles required to pump blood around the body. An ECG is recorded over a series of cardiac cycles (heartbeats) and shows the different phases of the cardiac cycle. The ECG indirectly measures transmembrane voltages in myocardial cells that depolarize and repolarize within each cardiac cycle. These depolarization and repolarization events produce ionic currents within the body, and these are transduced into voltages by electrodes (described in Chapter 2) placed on the surface of the chest and thorax, as shown in Figures 5.1(a) and (b). Up to twelve different lead voltages are recorded, with the magnitude of the voltages being in the low mV range, Figure 5.1(c), and a frequency spectrum between 0 and 30 Hz, as shown in Figure 5.1(d). The ECG signal has many distinct features, such as the P-wave, QRS-complex and T-wave, illustrated in Figure 5.1(c). The amplitude, shape and relative timing of these features can be used to diagnose different clinical conditions.

An ECG is an essential part of diagnosing and treating patients with acute coronary syndromes and is the most accurate method of diagnosing ventricular conduction disturbances and cardiac arrhythmias. It is also used to diagnose heart conditions such as myocardial infarcts, atrial enlargements, ventricular hypertrophies and blocks of the various bundle branches. An ECG is universally used to monitor a patient's cardiac activity during surgery.

Most ECG machines are now digital and automated, meaning that the data is analyzed automatically. Software algorithms measure different aspects (such as delays, durations and slopes) of the ECG waveform and provide a set of keyword interpretations of the scan such as 'abnormal ECG' or more specific suggested diagnoses such as 'possible sinoatrial malfunction'. The most common software package is the University of Glasgow algorithm, which is outlined in detail in section 5.4.1. The major interference signals are caused by electromagnetic coupling from external sources such as power lines, which must be removed by appropriate filtering.

5.1   Electrical Activity in the Heart

The structure of the heart is shown in Figure 5.2(a), together with the major coronary vessels and indications of the direction of blood flow. The myocardium (heart wall) is composed of cardiac muscle, which differs from skeletal muscle in

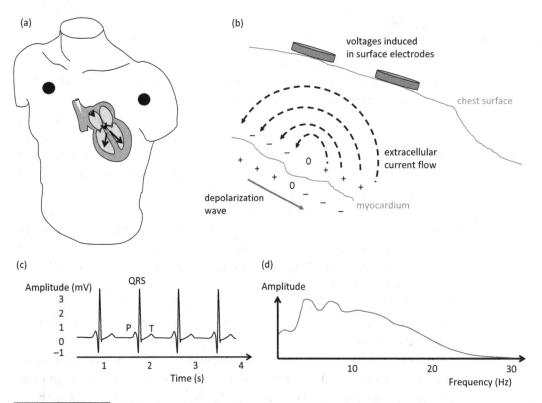

Figure 5.1 (a) A schematic illustration of the passage of cellular depolarization, which sweeps from the sinoatrial node through the heart in a top-left to bottom-right direction. (b) Extracellular current flow is transduced into a voltage recorded by electrodes placed on the body surface. (c) A typical ECG trace showing the P-wave, QRS-complex and T-wave. (d) A frequency spectrum of an ECG signal showing that the majority of the frequency components are below 30 Hz.

that skeletal muscles are voluntary (i.e. a person has complete control of their movement), whereas cardiac muscles are involuntary (i.e. the muscles contract and relax 'automatically', without conscious intervention). There are four compartments within the heart: the right and left ventricles, and the right and left atria. The two ventricles act as one subunit from an electrical activity point of view, with the two atria acting together as a separate subunit. The heart has four valves. The tricuspid valve lies between the right atrium and right ventricle, the mitral valve between the left atrium and left ventricle, the pulmonary valve between the right ventricle and pulmonary artery and the aortic valve in the outflow tract of the left ventricle controlling blood flow to the aorta.

In order to pump blood around the body the cardiac muscles must expand and contract in a specific order with well-defined timings between these coordinated actions. Muscle activity and timing are controlled by the propagation of cellular

(a)

(b)

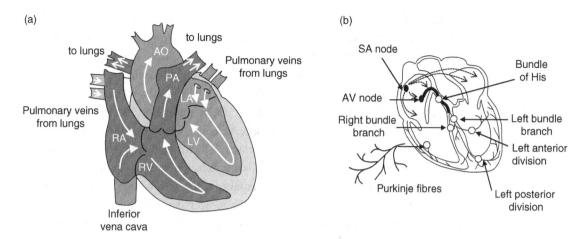

Figure 5.2   (a) Blood flow in the heart. (b) Electrical conduction in the heart.

activation through the cardiac system, as shown in Figure 5.2(b). At rest the cell membrane is impermeable to Na^+ and Ca^{2+}, but K^+ can cross the membrane freely via a channel that is specific to potassium ions. The intracellular concentrations are 15 mM Na^+, 10^{-4} mM Ca^{2+} and 150 mM K^+, and extracellular concentrations 145 mM Na^+, 1.8 mM Ca^{2+} and 4.5 mM K^+. Electrical activation in the muscle cells (myocytes) occurs via the inflow of sodium ions across the cell membrane.

The initiation point for the spreading action potential is the sinoatrial (SA) node, which contains specialized muscle cells that are self-excitatory, i.e. they need no external trigger: these are referred to as **pacemaker cells**. Pacemaker cells produce an action potential, which spreads throughout both atria. The atrioventricular (AV) node is situated at the boundary between the atria and ventricles, and is the focus of the conducting path from the atria to the ventricles. This conduction path consists of the bundle of His, which separates distally into two bundle branches that propagate along each side of the septum. The bundles later branch into Purkinkje fibres that diverge to the inner sides of the ventricular walls. At the AV node, electrical conduction is slowed and 'paused' for a short period of time to allow blood from the atria to fill the ventricles. Once the conduction signal passes through the AV node, the depolarization wavefront propagates rapidly towards the outer wall of the ventricle. Various diseases of the heart disrupt this electrical conduction pathway. For example, some diseases cause a delay in the passage through the AV node, meaning that ventricular repolarization is delayed with respect to a healthy heart.

There are two types of action potentials, slow and fast. Slow-response action potentials are characteristic of AV and SA node cells. Compared to cells with fast action potentials, depolarization is slower, no plateau phase exists and there is no true resting membrane potential. Na^+ channels do not contribute to the action potential, and instead depolarization of the cells is mediated by the opening of Ca^{2+} channels that allow inward flow of Ca^{2+}. Cells repolarize slowly through the opening of K^+ channels. In slow-response cells the action potentials have a maximum negative potential of approximately –65 mV and then slowly depolarize *spontaneously*. Since the slow-acting cells exhibit spontaneous depolarization they can activate repetitively and act as pacemaker cells. This allows for spontaneous and repetitive activation of the heart. The slower depolarization upstroke also means that these cells conduct impulses less rapidly. Slow conduction properties of cells in the AV node allow a temporal delay, which coordinates atrial and ventricular contraction, and also protect the ventricles from any rapid atrial arrhythmias.

Atrial and ventricular myocytes exhibit fast-response action potentials. In fast-response action potentials, depolarization of the cell membrane leads to opening of specialized Na^+ channels, allowing Na^+ to flow rapidly into the cell. An action potential of ~100 mV occurs for ~300 ms, after which cells repolarize via the outflow of potassium ions, as shown in Figure 2.4 in Chapter 2.

The electrical activity of the heart is shown as a series of time snapshots in Figure 5.3. The ionic currents that flow through the body fluids are created by

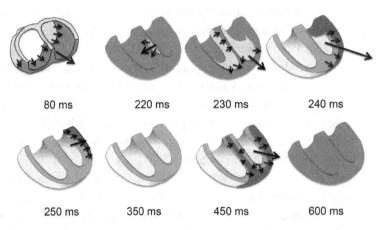

80 ms	220 ms	230 ms	240 ms

250 ms	350 ms	450 ms	600 ms

Figure 5.3 Successive steps in depolarization (light areas) and repolarization (dark areas) of the heart. Atrial depolarization ~80 ms, septal depolarization ~220 ms, apical depolarization ~230 ms, left ventricular depolarization ~240 ms, late left ventricular depolarization ~250 ms, ventricles completely depolarized ~350 ms, beginning of ventricular repolarization ~450 ms and ventricles completely repolarized ~600 ms. The thick black arrow shows the direction of the cardiac vector.

electric dipoles formed at the border of the depolarized and polarized cardiac tissue. The direction of the current is from the depolarized (positive surface) to the repolarized (negative surface), as shown in Figure 5.1(c). The **cardiac vector** indicates the overall direction of depolarization, i.e. the summation of the individual depolarization vectors as shown in Figure 5.3. As the depolarization wave sweeps across the heart, the cardiac vector changes in direction and magnitude.

5.2 Electrode Design and Einthoven's Triangle

As covered in Chapter 2, surface electrodes convert ionic current flow through the body into a potential, or potential difference between two electrodes, at the skin surface. A silver/silver chloride electrode is used, with a high conductivity NaCl gel forming the interface between the electrode and the skin surface. The skin is typically prepared by cleaning with alcohol and gentle abrasion to remove the upper skin layer, which has a very high impedance.

The simplest geometrical recording arrangement uses four electrodes: three active and one ground. The three active electrodes are placed on the left arm (LA), right arm (RA) and left leg (LL). The ground electrode is placed on the right leg (RL) and is attached to the ground of the ECG system. Pairs of electrodes are known as **leads**, and the potential differences between electrode pairs are known as **lead voltages**. There are three leads and three lead voltages (V_I, V_{II} and V_{III}) from the three active electrodes, which are defined as:

$$\text{Lead I}: \quad V_I = V_{LA} - V_{RA}$$
$$\text{Lead II}: \quad V_{II} = V_{LL} - V_{RA}$$
$$\text{Lead III}: \quad V_{III} = V_{LL} - V_{LA}$$

Net current flow towards the first electrode in the pair is defined as a positive voltage in the recorded waveform, and vice versa. As can easily be seen:

$$\text{Lead II} = \text{Lead I} + \text{Lead III} \tag{5.1}$$

This is known as Einthoven's law, i.e. the three standard limb leads contain only two pieces of independent information. Leads I, II and III are referred to as **bipolar leads**, since the voltages are measured as potential differences between two active electrodes. Figure 5.4(a) shows the set up for Einthoven's triangle. The voltage measured in a lead represents the projection of the cardiac vector onto the lead direction. Figure 5.4(b) shows two representations of the electrical activity of the heart, with the corresponding ECG trace recorded by Lead I.

Figure 5.4

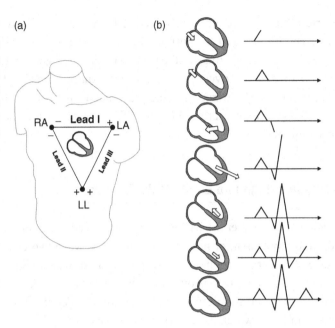

(a) Illustration of the three leads of Einthoven's triangle surrounding the heart. (b) The changing direction and magnitude of the total cardiac vector produces the ECG trace in Lead I.

Figure 5.5 illustrates how the recorded ECG trace comprises the summation of the individual traces recorded from different areas of the heart. Notice the different waveforms from the SA and AV node (relatively slow depolarization) vs. those from the other cardiac areas (rapid depolarization). Figure 5.5 shows six specific peaks, referred to as P, Q, R, S, T and U. The P-wave is produced by atrial depolarization. The Q, R and S-waves are produced at different stages of ventricular depolarization, and together form what is termed the QRS-complex. The duration of the complex is measured from the onset of the negative Q-wave to the termination of the negative S-wave. The T-wave is produced by ventricular repolarization. Since repolarization is slower than ventricular depolarization the T-wave is longer and has a lower amplitude than the QRS-complex. The U-wave is small, and often not seen, and is related to the repolarization of the Purkinje fibres. In addition to the individual peaks, the J-point refers to the isoelectric point at the end of the S-wave, when the entire myocardium is depolarized and is at zero potential.

5.2.1 Standard Twelve-Lead Configuration

As can readily be appreciated, if only one lead voltage is recorded then whenever the cardiac vector is perpendicular to the particular lead direction then zero voltage is recorded and information is effectively lost. This is one reason why many lead

Figure 5.5

Illustration of the formation of the P, Q, R, S, T and U-waves, which make up the ECG trace via addition of individual action potentials from different locations within the heart. Also shown is the isoelectric J-point.

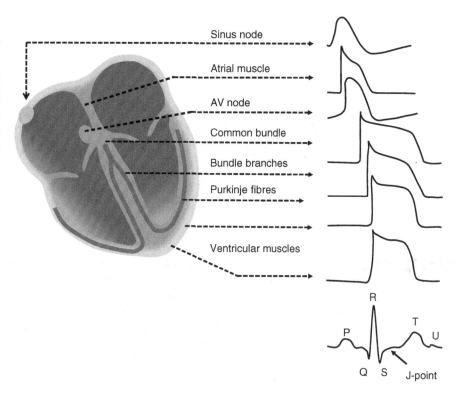

Figure 5.5

voltages are recorded. The full diagnostic ECG uses ten electrodes, which allows twelve different traces to be recorded. Figure 5.6 shows the placement of the six precordial electrodes (V1 to V6), which are added to the four (LA, LL, RA and RL) described previously, to make ten in total. A virtual reference terminal, called Wilson's central terminal (WCT), is defined as the average of the potential at the three limb electrodes:

$$V_{WCT} = (V_{LA} + V_{RA} + V_{LL})/3 \qquad (5.2)$$

The six precordial leads (V1 to V6) shown in Figure 5.6 are defined with respect to the WCT, and therefore represent **unipolar** leads, in contrast to the bipolar leads I, II and III. The general direction of ventricular activation in the frontal plane is called the cardiac axis. In a healthy person, the cardiac axis can range from −30° to +110°. This normal variation can be due to the varying position and orientation of the heart within the body. For example, thinner people tend to have a more vertical cardiac axis (+60° to +110°) because the ventricles are oriented more downwards.

Augmented right, left and front leads (aVR, aVL and aVF, respectively) can be created by defining Leads I, II and III with respect to the Goldberger central

(a)

(b)

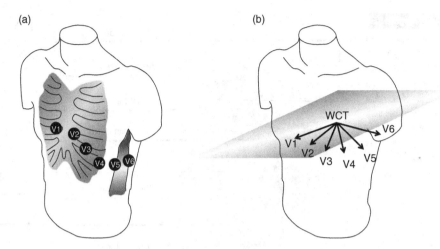

Figure 5.6 (a) Placement of the six precordial electrodes. Electrode V1 should be placed on the 4th intercostal space to the right of the sternum, electrode V2 on the 4th intercostal space to the left of the sternum, electrode V3 midway between electrodes V2 and V4, electrode V4 on the 5th intercostal space at the midclavicular line, electrode V5 on the anterior axillary line at the same level as electrode V4, and electrode V6 at the midaxillary line at the same level as electrodes V4 and V5. (b) Corresponding lead voltages V1 through V6, measured with respect to Wilson's central terminal.

terminals of $(LA + LL)/2$ for aVR, $(RA + LL)/2$ for aVL, and $(RA + LA)/2$ for aVF. This results in:

$$aVR = -\frac{\text{lead I} + \text{lead II}}{2}; \quad aVL = \text{lead I} - \frac{\text{lead II}}{2}; \quad aVF = \text{lead II} - \frac{\text{lead I}}{2}$$

(5.3)

These unipolar augmented leads have a 50% higher signal strength than the bipolar leads, but it must be emphasized that these values are *derived* from Leads I, II and III, and are not independent measurements. The standard 12-lead ECG consists of the three limb leads (Leads I, II and III), three augmented limb leads and the six independent precordial leads. These are shown in Table 5.1.

A typical plot of the twelve different leads is shown in Figure 5.7. For the standard digital display four 2.5 s-long columns are presented sequentially. The first column contains three rows representing Leads I, II and III; the second column contains three rows representing aVR, aVL and aVF; the third column displays Leads V1, V2 and V3; and the fourth column Leads V4, V5 and V6. At the bottom of the trace, Lead II is displayed for the full 10 seconds. This bottom plot is what is normally referred to as 'the ECG trace'.

Table 5.1 Lead definitions, actual or derived, and displayed or non-displayed.

Lead I (bipolar)	LA–RA	Actual	Displayed on monitor
Lead II (bipolar)	LL–RA	Actual	Displayed on monitor
Lead III (bipolar)	LL–LA	Derived	Displayed on monitor
V_{WCT} **(Wilson)**	1/3(LA + RA + LL)	Actual	Not displayed
aVR (unipolar)	$-\frac{1}{2}$ (Lead I + Lead II)	Derived	Displayed on monitor
aVL (unipolar)	Lead I $- \frac{1}{2}$ Lead II	Derived	Displayed on monitor
aVF (unipolar)	Lead II $- \frac{1}{2}$ Lead I	Derived	Displayed on monitor
V1 (unipolar)	$VC1 - V_{WCT}$	Actual	Displayed on monitor
V2 (unipolar)	$VC2 - V_{WCT}$	Actual	Displayed on monitor
V3 (unipolar)	$VC3 - V_{WCT}$	Actual	Displayed on monitor
V4 (unipolar)	$VC4 - V_{WCT}$	Actual	Displayed on monitor
V5 (unipolar)	$VC5 - V_{WCT}$	Actual	Displayed on monitor
V6 (unipolar)	$VC6 - V_{WCT}$	Actual	Displayed on monitor

Figure 5.7

Typical ECG display of twelve different leads, and an extended time recording of Lead II at the bottom.

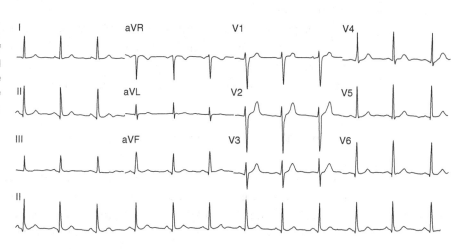

5.3   ECG System Design

The Committee on Electrocardiography of the American Heart Association, and many other similar organizations, has made specific recommendations for standard requirements of an ECG system. These include both measurement and safety specifications.

Measurement:

1. The instrument should be able to measure signals in the range 0.05 to 10 mV (the normal ECG signal is ~2 mV).
2. The input impedance of an electrode should be greater than 5 MΩ measured at a frequency of 10 Hz.

3. The instrument should have a measurement bandwidth from 0.1 to 150 Hz.
4. Over the measurement bandwidth there should be a phase variation less than that of an analogue 0.05 Hz single-pole high-pass filter to ensure that there are no distortions of the T-wave and ST segments of the ECG waveform.
5. The instrument should have a very high common-mode rejection ratio (>80 dB) before the first amplification stage.

Safety:
6. The instrument should not allow leakage currents (see Chapter 9) greater than 10 µA to flow through the patient.
7. Isolation methods should be used to keep the patient from becoming part of an AC electrical circuit in the case of a patient-to-powerline fault.

The following sections show how the ECG system can be designed to fulfil these criteria. Figure 5.8 shows the general outline of an ECG instrument. The first block is patient protection and defibrillation pulse clamping. This block comprises high-value resistors or any other kind of isolation circuitry. The lead selection circuitry determines the various electrode combinations to be measured. Design of the remaining modules is covered in the following sections.

5.3.1   Common-Mode Signals and Other Noise Sources

As discussed in the first section of this chapter the magnitude of the measured ECG biopotential is a few mV, covering a frequency range from DC to roughly 30 Hz. However, if one looks at the actual lead voltages recorded, they appear as in Figure 5.9. There is a very large offset voltage, much larger than the actual AC voltage that represents the ECG signal. This offset voltage has two main components: the first is from common-mode 50/60 Hz interference, and the second from the DC electrode offset potential: as outlined in Chapter 2 this is due to the half-cell potential developed at the electrodes, which has a value of approximately +/–300 mV for the silver/silver chloride system used for most ECG electrodes.

Figure 5.10 shows different paths though which common-mode voltages can occur. First, the body itself can act as a receiving antenna via capacitive coupling to the mains power line. The ECG cables can also act as antennas, and pick up capacitively coupled signals from the power lines.

Other sources of noise include the following:

(i) Electrode pop or contact noise: any loss of contact between the electrode and the skin giving a sharp change in signal amplitude.

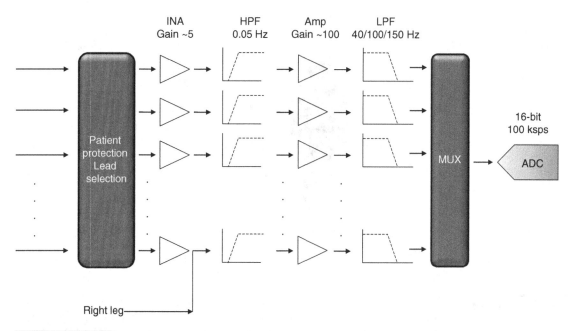

Figure 5.8 Block diagram of the ECG system. The instrumentation amplifier (INA) provides low first-stage gain. The signals from each electrode are typically time-domain multiplexed (MUX) into a single ADC.

Figure 5.9

Schematic showing the components of the ECG trace data (not to scale), which appears as a small time-varying voltage superimposed on top of a much larger constant offset voltage from the electrodes and a 50/60 Hz common-mode signal from coupling to the power lines.

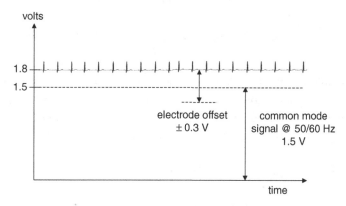

(ii) Patient–electrode motion artefacts: any displacement of the electrode away from the contact area on the skin leads to variations in the impedance between the electrode and skin causing rapid signal and baseline jumps.

(iii) EMG interference: electrical activity due to muscle contractions gives signals that last for approximately 50 ms and occur at frequencies between DC and 10 kHz with an average amplitude of ~10% of that of the ECG signal.

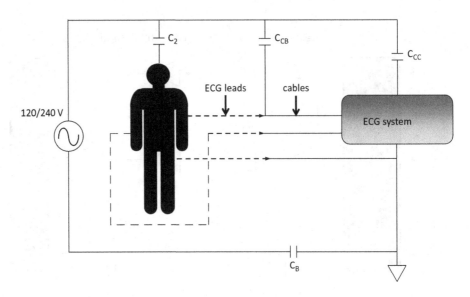

Schematic of some of the different paths by which a common-mode voltage can be coupled into the body via the skin surface. C_2 represents capacitive coupling between the mains power line and the body, C_{CB} coupling between the power line and the ECG leads, C_B represents the coupling between the ground of the ECG system and that of the hospital power supply, and C_{CC} coupling between the ECG system itself and the power lines.

(iv) Baseline drift, from patient breathing at frequencies between 0.15 and 0.3 Hz, has an amplitude of ~15% of that of the ECG signal.

(v) Noise generated by other medical equipment in the close vicinity.

As seen in Figure 5.8 the second stage of the ECG system after the electrodes is an amplifier, and if the large common-mode signal is not removed it would saturate this first-stage amplifier. Therefore it is important to try to reduce the common-mode signal as much as possible, as covered in the next section.

5.3.2   Reducing the Common-Mode Signal

As outlined previously, the dominant form of common-mode interference comes from the electrical power system coupling into the body and the ECG leads. Ideally, this common-mode signal would be identical for each lead and could therefore be removed by a differential amplifier. However, slight differences in the degree of coupling due to variations in the skin–electrode impedance of different leads means that subtraction is never perfect. An instrumentation amplifier, together with a driven right leg negative feedback circuit, can be used to reduce the common-mode signal so that the level is well below that of the ECG signal itself.

5.3.2.1   Instrumentation Amplifier

As covered in section 3.3.3, the common-mode differential signal can be reduced using a three-op-amp configuration known as an instrumentation amplifier. The common-mode signal is not completely eliminated since the coupled signals in two different ECG leads may not be exactly the same, and also the two input op-amps may not be exactly matched in terms of input impedance and gain.

5.3.2.2   Driven Right Leg Circuit

In order to increase the degree of cancellation of the common-mode signal, a fraction of this signal is inverted and driven back into the patient through an electrode placed on the right leg: this is referred to as a right leg drive (RLD) or driven right leg (DRL) circuit. A return current of only a few microamps or less is required to achieve significant improvement in the CMRR. As shown in Figure 5.11, the common-mode voltage on the body is sensed by the two averaging resistors, inverted, amplified (using an auxiliary op-amp) and fed back into the right-leg electrode. This essentially forms a negative feedback circuit, driving the common-mode voltage close to zero. This feedback loop improves the common-mode rejection by an amount equal to $(1 + A)$, where A is the closed-loop gain of the feedback loop. The value of the closed-loop gain is given by:

$$A = 2\frac{R_F}{R_{CM}(1 + j\omega R_F C_F)} \tag{5.4}$$

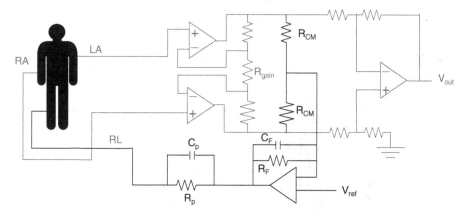

Figure 5.11   Full circuit diagram of the instrumentation amplifier and closed-loop RLD to reduce the common-mode signal. The instrumentation amplifier is shown in grey and the RLD circuit in black. R_p is a protection resistor to limit the current fed back into the patient, and the parallel C_p ensures that the feedback loop is stable.

This circuit also increases the patient's safety. If electrical leakage were to cause a high voltage to appear between the patient and ground, the auxiliary op-amp on the right leg would saturate, which effectively ungrounds the patient.

5.3.3   Design of Lead-Off Circuitry

Ensuring that the electrodes are all securely in good contact with the body is essential for obtaining high-quality ECG signals. Over time, the conductive path between the patient and a gelled electrode can break down or dry out, causing an increased impedance of the skin-electrode interface from a few kΩ to well over 10 kΩ over a period of hours.

In order to detect this effect, a 'lead-off' circuit is incorporated into the ECG design. A very low amplitude low-frequency (typically DC or 30 kHz) current is applied to the patient, as shown in Figure 5.12. DC-based lead-off detection requires the ECG system to include a return path for this excitation signal, which can come from the RLD connection. If the skin-electrode impedance increases due to a reduced degree of contact, the current that enters the amplifier (as opposed to the body) increases, thereby increasing the DC offset and ultimately causing the amplifier to saturate. An analogue comparator is used to compare the input voltage to a user-set voltage level designed to sound an alert when the impedance between the patient and electrode exceeds the ECG system design specifications. The disadvantages of the DC method are that there is an additional offset voltage on the ECG signal, and also some additional noise, both of which need to be filtered out. The alternative AC method injects a low-frequency signal, which is then recorded in addition to the ECG signal. Since the frequency of this extra signal is known, it can be removed using a digital notch filter after the data

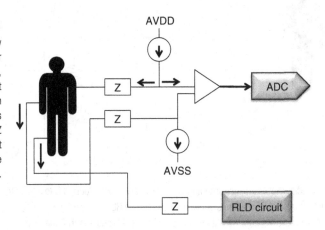

Figure 5.12

Schematic of the circuitry used to detect whether there is a lead-off event, i.e. the electrode contact with the body has been reduced. The arrows indicate current flow. Z represents the input impedance of the electrodes.

has been digitized. If the amplitude of this signal is above a predefined level then a lead-off event is indicated and an alarm sounds on the ECG console.

5.3.4 Filtering and Sampling

As shown in Figure 5.8 the total gain of the system is distributed between the instrumentation amplifier (INA) and second amplifier placed after the high-pass filter. The electrode voltage offset may be up to 300 mV, and since the supply voltage to the INA is typically 5 volts, the maximum INA gain is in the range of 5 to 10 to avoid saturation. After the INA, the DC offset is removed using a high-pass filter with a cut-off frequency of 0.05 Hz. The value of the cut-off frequency is chosen so that the level of the ST segment of the ECG is not distorted, since the ST segment level and shape are used in the diagnosis of many conditions (see section 5.5). After the DC component has been removed, the signal is further amplified by a factor of approximately 100. This signal next passes through a low-pass filter in order to remove out-of-band noise. Typically, a fourth- or higher order active low-pass filter is used. On clinical ECG systems, these low-pass filters have cut-off frequencies of 40, 100 or 150 Hz, the choice of which one to use being up to the operator. The standard recommendations are 100 Hz for adults and 150 Hz for children. However, if the ECG has a very strong contamination from muscle motion artefacts or other movements then the 40 Hz filter can be applied. Applying this very low cut-off frequency has the disadvantage that it significantly distorts sharp features in the ECG signal such as the top of the QRS-complex, which reduces its apparent amplitude. The LPF is followed by a MUX that feeds into the ADC. The sampling frequency should be >500 Hz so that at least ten samples are acquired over the QRS-complex, which typically has a duration of ~20 ms. Although the instrumentation amplifier and RLD system remove most of the power-line interference, some may still get through, and this is filtered using a digital notch filter at either 50 or 60 Hz, depending upon the particular country.

In very new ECG systems, much of the analogue filtering shown in Figure 5.8 can be replaced by post-acquisition digital signal processing, resulting in the simplified system shown in Figure 5.13. The same ECG front-end can be used, but in this case it is interfaced with a high resolution $\Delta\Sigma$ ADC, which uses the principles of oversampling and noise-shaping described in Chapter 4, with sampling rates up to hundreds of kilohertz. Figure 5.13 shows that the analogue high-pass filter, DC blocking filter, gain stage and active low-pass filter can be eliminated. As also covered in Chapter 4, a simple single-pole analogue RC filter can be used with the oversampling ADC, since its frequency characteristics are not critical.

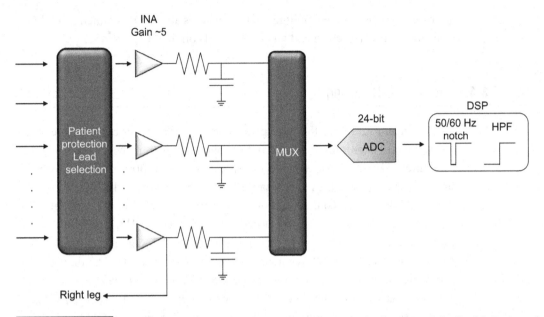

Figure 5.13 A simplified ECG set up in which most of the signal processing is performed after the data has been digitized by a high-resolution $\Delta\Sigma$ ADC.

5.4   Signal Processing of the ECG Signal and Automatic Clinical Diagnosis

The general shape of the ECG signal, as discussed earlier, consists of distinct sections. The P-wave represents the atrial depolarization, the QRS-complex the depolarization of the right and left ventricles, and the T-wave the repolarization of the ventricles. Within the QRS-complex the Q-wave is generated when the depolarization wave travels through the interventricular septum. The R-wave corresponds to a contraction of the left ventricle to pump blood out of the ventricle, and the S-wave to the basal sections of the ventricles depolarizing due to the contraction of the right ventricle. By analyzing the delays between, and amplitudes of, these sections and their reproducibility over time, a trained physician can relate abnormalities or distortions to specific pathological conditions. Table 5.2 lists typical characteristics of the ECG signal in healthy adults.

One of the things to note is that Table 5.2 reports values for a heart rate (HR) of 60 bpm. However, it is known that the QT interval, for example, is a function of the HR: normally the faster the HR the shorter the QT interval, and vice versa. For this reason, numerous clinical investigators have attempted to 'correct' the QT interval to a corrected value QTc, which would have been obtained if the HR had been 60 bpm. The three most common corrections are:

Table 5.2 Typical values for timings and amplitudes of different features of the ECG signal from Lead II for a heart rate of ~60 beats per minute for a healthy male.

	Normal value
P-wave width	110 ± 20 ms
P-wave amplitude	0.15 ± 0.05 mV
PR interval	160 ± 40 ms
QRS-complex width	100 ± 20 ms
R-wave amplitude	1.5 ± 0.5 mV
QTc interval	400 ± 40 ms
ST segment	100 ± 20 ms
T-wave duration	160 ± 20 ms
T-wave amplitude	0.3 ± 0.2 mV

$$\text{Bazett}: \quad QT_{cB} = QT(RR)^{-\frac{1}{2}}$$

$$\text{Fridericia}: \quad QT_{cFri} = QT(RR)^{-\frac{1}{3}}$$

$$\text{Framingham}: \quad QT_{cFr} = QT + 0.154(1 - RR) \tag{5.5}$$

5.4.1 University of Glasgow (Formerly Glasgow Royal Infirmary) Algorithm

The most common software algorithm/package used to diagnose common ECG abnormalities, such as myocardial infarction (anterior, inferior or a combination), ventricular hypertrophy (both left and right) and myocardial ischaemia, is the University of Glasgow, formerly Glasgow Royal Infirmary (GRI), algorithm that provides an interpretation of the resting 12-lead ECG. Conduction defects such as right bundle branch block (RBBB) or left bundle branch block (LBBB) and other abnormalities such as prolonged QT interval can also be detected. The Glasgow programme can be used in both adults and children. Figure 5.14 shows a schematic of a processed ECG signal with annotated features that form the inputs to the University of Glasgow algorithm.

There are many processing steps required to produce an output suitable for such an algorithm. Typically, a number (five to ten) of traces are averaged, with time-locking between the traces performed using cross-correlation techniques covered in Chapter 4: this produces more accurate and reproducible results than analyzing the waveform from a single cycle. In addition, the time coherence of simultaneously measured complexes can be used to derive global measurements of reported intervals such as those listed in Table 5.2. Waveform recognition is then performed based on the individual waveforms and delays between the waveforms.

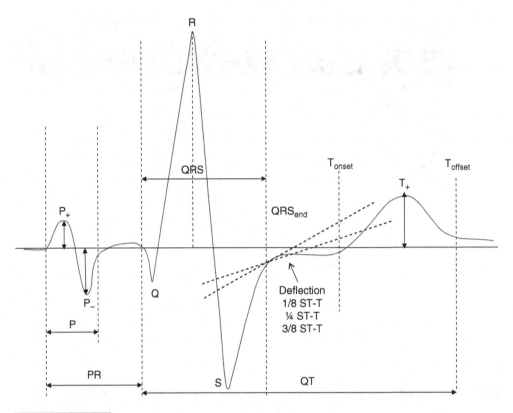

Figure 5.14   Diagram showing some of the parameters that are used as inputs to the University of Glasgow algorithm for automatic analysis of ECG waveforms.

5.5   Examples of Abnormal ECG Recordings and Clinical Interpretation

The features of an ECG recording give detailed information about possible pathological conditions and causes. A selection of the differences between an abnormal and normal ECG signal is given below.

The P-Wave is Inverted

There are a number of different possible causes. One is that atrial polarization occurs at a different physical location than normal, producing a phase change in the signal. Another is a sinoatrial block in which the pacemaker signal comes from another part of the heart, rather than the SA node.

The Magnitude of the P-Wave is Much Higher or Lower Than Usual

A much larger P-wave can indicate enlargement of the right atrium (see Problem 5.4). A much lower P-wave can be caused by a sinoatrial block, meaning that the SA node does not fire, and the pacemaker function is taken over by the AV junction. The P-wave can also disappear, being replaced by random electrical activity, in the case of atrial fibrillation shown in Figure 5.15(a): this causes the normal pacemaker potentials from the SA node to be much smaller than asynchronous impulses that originate in the atria.

The P-Wave is Elongated to Longer Than 100 ms and May Have an Additional Peak

This can indicate enlargement of the left atrium (see Figure 5.15(b) and Problem 5.5).

The Delay Between the P-Wave and R-Wave is Altered

A much longer delay (>120 ms) than normal is indicative of an impairment of the conduction path between the atria and ventricles. One potential cause is first-degree AV nodal block, shown in Figure 5.15(c). Third-degree AV nodal block involves the atria and ventricles essentially working independently of each other,

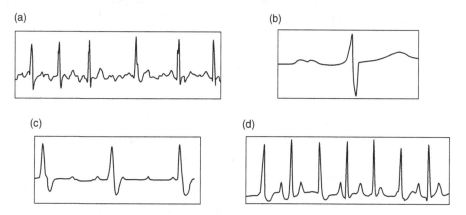

Figure 5.15   Examples of abnormal ECG traces. (a) Atrial fibrillations characterized by the P-wave disappearing. (b) P-mitrale produces an elongated P-wave with a notch in the signal. (c) First-order AV node block elongates the PR delay. (d) Third-degree AV nodal blockage decouples the P-wave and QRS-complex in terms of their time relationship.

and therefore the P-wave and the QRS-complex are decoupled in terms of the timing, as seen in Figure 5.15(d). In contrast, if the delay between the P-wave and R-wave is much shorter than 120 ms, this can be caused by conditions such as Wolff–Parkinson–White (WPW) syndrome in which a *secondary* conduction pathway leads to pre-excitation of the ventricles.

Change in the Vector Orientation of Maximum QRS Deflection

Abnormalities can also be detected by looking at the mean frontal plane electrical axis during the QRS-complex, i.e. the vector of the maximum QRS deflection. This value in healthy adults is between −30° and +110°. Moderate left-axis deviation is defined as being between −30° and −45°, and marked deviation between −45° and −90°, which is often associated with left anterior fascicular block (see Problems).

The ST Segment Amplitude is Higher (Elevated) or Lower (Depressed) Than Normal

The delay between the S- and T-wave, the ST segment, is the time required for the ventricles to pump the blood to the lungs and rest of the body. The most important cause of ST segment abnormality (elevation or depression) is myocardial ischaemia or infarction. If the level is depressed it can indicate ischaemia, resulting in a decrease in blood supply caused by obstructions in the blood vessels. In contrast, an elevation can indicate myocardial infarction, with heart tissue being damaged.

Abnormal Shape of the T-Wave

The T-wave is normally slightly asymmetric with the first half having a more gradual slope than the second half. If the T-wave is inverted it can indicate ischaemia. If it is tall and peaked it is indicative of hyperkalaemia, i.e. a very high concentration of potassium ions in the blood. If it is flat, this can indicate the opposite condition of hypokalaemia (see Problem 5.6).

There are a large number of accepted textual outputs in terms of the 'primary statements' that originate from an automatic analysis of the ECG waveforms. A brief summary is given in Table 5.3; full details of the 117 recommended statements can be found in reference [2]. Secondary statements suggest a

Table 5.3 Selected primary statements from an ECG output

Overall interpretation	Sinus node rhythms and arrhythmias	ST-segment, T-wave and U-wave
Normal ECG	Sinus rhythm	ST deviation
Otherwise normal ECG	Sinus tachycardia	ST deviation with T-wave change
Abnormal ECG	Sinus bradycardia	T-wave abnormality
Uninterpretable ECG	Sinus arrhythmia	Prolonged QT interval
	Sinoatrial block, type I	Short QT interval
Technical conditions	Sinoatrial block, type II	Prominent U waves
Extremity electrode reversal	Sinus pause of arrest	Inverted U waves
Misplaced precordial electrodes	Uncertain supraventricular rhythm	TU fusion
Missing leads		ST–T change due to ventricular hypertrophy
Artefact		Early repolarization
Poor-quality data		

tentative diagnosis such as: suggests acute pericarditis, acute pulmonary embolism, chronic pulmonary disease, hyperkalemia, hypothermia or sinoatrial disorder.

5.6 ECG Acquisition During Exercise: Detection of Myocardial Ischaemia

In addition to acquiring an ECG when the patient is at rest an ECG is often also acquired during exercise. Exercise stresses the heart, and so areas of infarct within the heart that cannot respond to the increased need for oxygen become more apparent than at rest. Data from exercise ECGs can be used for either diagnostic or prognostic purposes in patients with:

  (i) symptoms that suggest myocardial ischaemia
 (ii) known coronary artery disease with a recent check-up suggesting a change in clinical status
(iii) valvular heart disease, or
 (iv) recently-diagnosed heart failure or cardiomyopathy.

The general procedure is that a resting-state ECG is acquired first with the patient lying down and then standing up. Then a continuous series of ECGs is acquired during exercise (either on a treadmill or a stationary bicycle). Further recordings are made during the post-exercise recovery state (every two minutes for seven to ten minutes until the heart rate falls below 100 beats per minute or the ECG waveform returns to the control baseline pattern). A final ECG is acquired under fully relaxed conditions. The most common exercise protocol (the Bruce protocol) is divided into successive three-minute stages, each of which requires the patient to walk faster and at a steeper grade. This procedure is not without risks. Serious complications such as acute myocardial infarction occur approximately once for every 10 000 tests performed.

As an illustration of the information that can be derived from an ECG acquired during exercise, two common conditions are considered here: transmural and subepicardial cardiac infarcts, as represented in Figure 5.16. The most relevant ECG feature studied is ST-segment deviation, which is reported as ST-elevation or ST-depression. In fact, as explained below, these are quite misleading terms if taken literally, since the apparent elevation or depression of the ST-segment actually reflect physiological conditions that result in different baselines for the ECG signal.

Changes in the ST-segment can be understood in terms of a **current-of-injury mechanism** shown in Figure 5.16. In a healthy person, the ST segment is usually nearly isoelectric (i.e. flat along the baseline) because healthy myocardial cells attain approximately the same potential during early repolarization, which corresponds to the plateau phase of the ventricular action potential. However, as a result of the decreased O_2 perfusion in ischaemic tissue, there is decreased production of ATP, which is required for the final stages of cellular repolarization from a membrane potential of -80 mV to -90 mV. Therefore ischaemic cells repolarize more slowly than healthy ones, and still retain a degree of depolarization during the TP segment. The difference in membrane potential between healthy and ischaemic cells causes negative charges to accumulate on their surfaces, generating a cardiac vector that points towards the healthy cardiac cells.

In the case of transmural injury that affects the entire myocardial wall, shown in Figure 5.16(a), the cardiac vector points away from the chest ECG leads, causing a

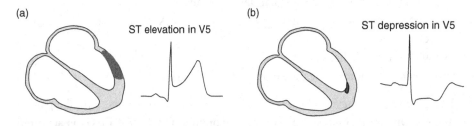

Figure 5.16   Current-of-injury patterns in myocardial ischaemia. (a) Ischaemia involving the entire ventricular wall (transmural injury) results in a cardiac vector that produces a negative signal in the pericardial leads during the TP segment, with associated elevation of the ST segment. (b) In cases of subendocardial ischaemia, the cardiac vector produces a positive signal in the leads during the TP segment, resulting in an apparent depression in the ST segment.

downward deflection in the TP segment. However, since the TP segment forms the baseline of the ECG, the machine corrects for this by raising the TP segment to form the baseline, which results in elevation of the ST segment. In contrast, when ischaemia is confined primarily to the subendocardium, as shown in Figure 5.16 (b), the overall ST vector typically shifts towards the inner ventricular layer and the ventricular cavity such that the anterior precordial leads now show ST segment depression, again due to the ECG system assigning a baseline to the signal during the TP segment.

5.7 High-Frequency (HF) ECG Analysis

As mentioned previously, the conventional frequency bandwidth used for detecting ECG signals is 0.05 to 150 Hz (see guidelines in section 5.3), with the vast majority of the signal found in the 0.05 to 30 Hz range. However, there are also much higher frequency components in the ECG signal, particularly within the QRS-complex [3]. It has been suggested that these very high frequency components have diagnostic value, although at the current time they have not been integrated into standard clinical care. In the case of myocardial ischaemia the frequency range 150 to 250 Hz has most often been studied, whereas in studies of previous myocardial infarction a frequency range of 80 to 300 Hz is more common [4]. The physiological causes of high-frequency voltage fluctuations have not yet been definitively established, but one theory is that the high-frequency components are related to the conduction velocity and the fragmentation of the depolarization wave in the myocardium [5].

Since the ECG signal is typically sampled at rates up to 500 Hz, high-frequency components can potentially be extracted from standard clinical measurements, although it would be better to sample at least two to three times faster and so most dedicated measurements for high-frequency ECG have sampling rates of 1 kHz or higher. These sampling rates can easily be achieved using high-resolution $\Delta\Sigma$ over-sampling ADCs as shown earlier in Figure 5.13. For high-frequency measurements signal averaging is necessary since the signals have amplitudes in the order of tens of microvolts, less than 1% of the conventional ECG signal. The high-frequency signals are band-pass filtered before digitization. The largest high-frequency components are normally found in Leads V2, V3 and V4, as well as II, aVF and III.

Figure 5.17(a) presents a typical example of the high-frequency QRS (HF-QRS) signal during different stages of an exercise stress test of an ischaemic patient. In this example the HF-QRS signal shows significant decreases in amplitude in multiple leads as the exercise test progresses, which is indicative of ischaemic heart disease. In contrast, the HF-QRS signal remains essentially unchanged for a healthy subject, as shown in Figure 5.17(b).

Figure 5.17

(a) An example of a standard ECG signal and HF-QRS signal as a function of heart rate for a patient with ischaemic heart disease undergoing an ECG test during exercise. (b) Corresponding plots for a healthy subject.

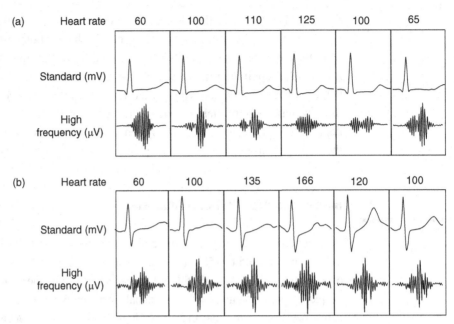

Several methods are used to quantify the HF-QRS signal. One of the most common is to calculate the RMS voltage (A) during the QRS-complex:

$$A = \sqrt{\frac{1}{n}\sum_{i=1}^{n} A_i^2} \qquad (5.6)$$

where n is the number of measurements, and A_i are the individual measured amplitudes. The standard ECG recording is used for determining exactly where the QRS-complex starts [6].

Another widely used quantification method is the detection of reduced amplitude zones (RAZs), which are characterized by a 'split' in the amplitude envelope of the HF-QRS-complex, as shown in Figure 5.18. A RAZ is defined when at least two local maxima or two local minima are present within one HF-QRS envelope, and when each local maximum or minimum is defined by its having an absolute voltage that is higher or lower than the three envelope sample points immediately preceding and following it [7]. In contrast, normal HF-QRS envelopes have a single local maximum and minimum. Three different types of RAZs have been identified: Abboud RAZ, Abboud Percent RAZ and NASA RAZ (the most severe) [8].

The detection of RAZs is used in particular in patients experiencing chest pain, and who have what is termed non-ST elevation myocardial infarction (NSTEMI). Despite the serious cardiac condition these patients often present

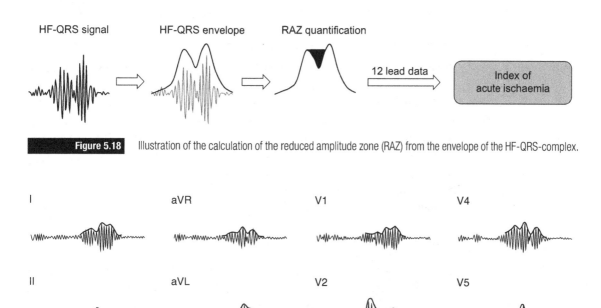

Figure 5.18   Illustration of the calculation of the reduced amplitude zone (RAZ) from the envelope of the HF-QRS-complex.

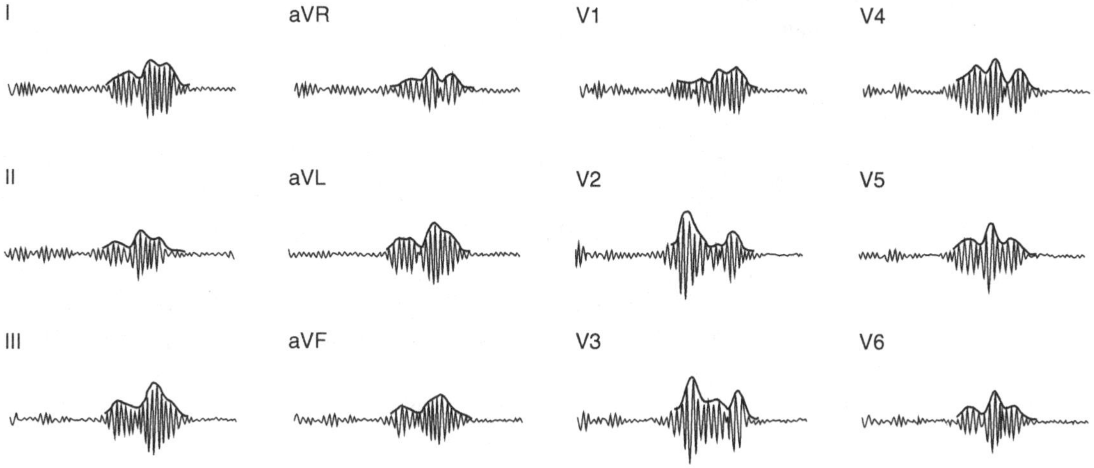

Figure 5.19   An example 12-lead high-frequency QRS signal from a patient with non-ST elevation myocardial infarction, showing RAZs in many of the leads.

with a normal ECG. Figure 5.19 shows results from an examination of such an NSTEMI patient in which six of the twelve leads show significant RAZs, which are automatically detected by specialized software.

Other HF signals which are potentially useful for diagnosis are the ventricular late potentials (VLPs), which occur typically at the end of the QRS-complex and within the ST-T segment. These are present in areas of the heart in which the conduction velocity is slower than in healthy tissue, and can be caused by ischaemia or collagen deposition after myocardial infarction. Although, as stated previously, not widely used in standard clinical practice there are an increasing number of clinical research reports about the diagnostic performance of HF-ECG including left ventricular hypertrophy, post heart-transplantation and in ischaemic heart disease in general [4, 8–17].

PROBLEMS

5.1 Describe the pumping action of the heart linking muscle contractions in the atria and ventricles and opening and closing of each of the valves.

5.2 If a drug blocks the K^+ channels in cardiac myocytes, what effect would this have on the action potentials?

5.3 If a drug blocks the Na^+ channels in cardiac myocytes, what effect would this have on the action potentials?

5.4 In Figure 5.4 plot the ECG trace for Lead II.

5.5 In Figure 5.4 plot the ECG trace for Lead III.

5.6 Suggest two different ways of estimating the cardiac axis from individual ECG lead recordings.

5.7 Explain why a high-input impedance amplifier is required in an ECG system, whereas the skin–electrode impedance should be as low as possible.

5.8 In Figure 5.8 design a single-pole high-pass filter, and a switchable second-order low-pass filter (switchable between 10, 100 and 150 Hz).

5.9 Calculate the phase variation for a HPF ($R = 318k\Omega$, $C = 10\mu F$) over the range of 1 to 150 Hz.

5.10 If the raw ECG signal is 3 mV and the common-mode signal is 1.5 volts, how large does the CMRR have to be to reduce the common-mode signal to less than the ECG signal?

5.11 In Figure 5.11 show that the closed-loop gain of the RLD circuit is given by:

$$A = 2\frac{Z_F}{R_{CM}} = 2\frac{R_F}{R_{CM}(1 + j\omega R_F C_F)}$$

5.12 Explain why a much larger P-wave can indicate enlargement of the right atrium.

5.13 Explain why the observation that the P-wave is elongated to longer than 100 ms and may have an additional peak, can indicate enlargement of the left atrium.

5.14 Explain why:

   (i) if the T-wave is inverted it can indicate ischaemia
   (ii) if it is tall and peaked, it is indicative of hyperkalaemia, and
   (iii) if it is flat, then this indicates hypokalaemia.

5.15 Why is the SNR of very high-frequency ECG signals shown in Figure 5.17 much lower than the conventionally recorded trace. Explain some of the problems associated with signal averaging, and techniques that might be used to overcome these problems.

5.16 Investigate the phenomenon of left anterior fascicular block (LAFB). Explain the altered conduction path through the heart, and relate this to the ECG traces that are recorded in different leads. Why does this condition result in the vector of the maximum QRS deflection being shifted from normal to between $-45°$ and $-90°$?

5.17 Any condition that places an abnormal load on the left ventricle can lead to left ventricular hypertrophy, i.e. the ventricle becomes larger and thicker. What features of the ECG trace would change based on these changes in size?

REFERENCES

[1] Kusumoto, F. *ECG Interpretation: From Pathophysiology to Clinical Applications*. New York: Springer; 2009.

[2] Mason, J. W., Hancock, E. W., Gettes, L. S. *et al.* Recommendations for the standardization and interpretation of the electrocardiogram. Part II: Electrocardiography diagnostic statement list. A scientific statement from the American Heart Association Electrocardiography and Arrhythmias Committee, Council on Clinical Cardiology; the American College of Cardiology Foundation; and the Heart Rhythm Society. *Heart Rhythm* 2007; **4**(3):413–19.

[3] Golden, D. P. Jr., Wolthuis, R. A. & Hoffler, G. W. A spectral analysis of the normal resting electrocardiogram. *IEEE Trans Biomed Eng* 1973; **20**(5):366–72.

[4] Tragardh, E. & Schlegel, T. T. High-frequency QRS electrocardiogram. *Clin Physiol Funct Imaging* 2007; **27**(4):197–204.

[5] Abboud, S., Berenfeld, O. & Sadeh, D. Simulation of high-resolution QRS complex using a ventricular model with a fractal conduction system. Effects of ischemia on high-frequency QRS potentials. *Circ Res* 1991; **68**(6):1751–60.

[6] Xue, Q., Reddy, B. R. & Aversano, T. Analysis of high-frequency signal-averaged ECG measurements. *J Electrocardiol* 1995; **28** Suppl:239–245.

[7] Abboud, S. Subtle alterations in the high-frequency QRS potentials during myocardial ischemia in dogs. *Comput Biomed Res* 1987; **20**(4):384–95.

[8] Schlegel, T. T., Kulecz, W. B., DePalma, J. L. *et al.* Real-time 12-lead high-frequency QRS electrocardiography for enhanced detection of myocardial ischemia and coronary artery disease. *Mayo Clin Proc* 2004; **79**(3):339–50.

[9] Conti, A., Alesi, A., Aspesi, G. *et al.* High-frequency QRS analysis compared to conventional ST-segment analysis in patients with chest pain and normal ECG referred for exercise tolerance test. *Cardiol J* 2015; **22**(2):141–9.

[10] Conti, A., Bianchi, S., Grifoni, C. *et al.* 2d.03: Improving diagnostic strategy in patients with long-standing hypertension, chest pain and normal resting ECG: value of the exercise high-frequency QRS versus ST-segment analysis. *J Hypertens* 2015; **33** Suppl 1:e28–29.

[11] Conti, A., Bianchi, S., Grifoni, C. *et al.* High-frequency QRS analysis superior to conventional ST-segment analysis of women with chest pain. *Am J Emerg Med* 2016; **34**(3):437–42.

[12] Amit, G., Granot, Y. & Abboud, S. Quantifying QRS changes during myocardial ischemia: insights from high frequency electrocardiography. *J Electrocardiol* 2014; **47**(4):505–11.

[13] Leinveber, P., Halamek, J. & Jurak, P. Ambulatory monitoring of myocardial ischemia in the 21st century: an opportunity for high-frequency QRS analysis. *J Electrocardiol* 2016; **49**(6): 902–6.

[14] Rahman, A. M., Gedevanishvili, A., Bungo, M. W. *et al.* Non-invasive detection of coronary artery disease by a newly developed high-frequency QRS electrocardiogram. *Physiol Meas* 2004; **25**(4):957–65.

[15] Rahman, M. A., Gedevanishvili, A., Birnbaum, Y. *et al.* High-frequency QRS electrocardiogram predicts perfusion defects during myocardial perfusion imaging. *J Electrocardiol* 2006; **39**(1):73–81.

[16] Spackman, T. N., Abel, M. D. & Schlegel, T. T. Twelve-lead high-frequency QRS electrocardiography during anesthesia in healthy subjects. *Anesth Analg* 2005; **100**(4):1043–7.

[17] Tragardh, E., Schlegel, T. T., Carlsson, M. *et al.* High-frequency electrocardiogram analysis in the ability to predict reversible perfusion defects during adenosine myocardial perfusion imaging. *J Electrocardiol* 2007; **40**(6):510–14.

6 Electroencephalography

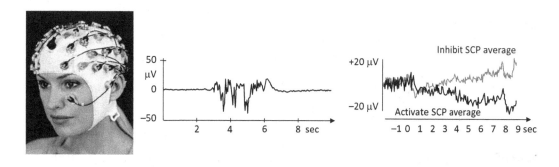

Introduction

An EEG measurement uses several electrodes placed on the scalp of the patient to measure indirectly the electrical activity of the brain, which consists of neuronal

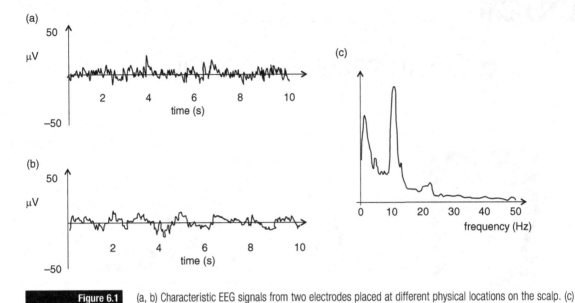

Figure 6.1 (a, b) Characteristic EEG signals from two electrodes placed at different physical locations on the scalp. (c) A frequency spectrum of EEG recordings summed from all areas of the scalp.

action potentials. Similar to ECG measurements considered in the previous chapter, EEG electrodes transduce the internal ionic currents conducted through the brain (acting as an inhomogeneous volume conductor) into potential differences that can be measured on the skin surface. Up to 32 electrodes are placed around the brain, with one electrode serving as a reference. Electrode design is very similar to that described in Chapter 2.

Typical EEG signals and their corresponding frequency spectra are shown in Figure 6.1. The recorded voltages are much smaller than those produced in an ECG, with peak-to-peak amplitudes less than 100 microvolts. The frequency content ranges from zero to ~50 Hz, with specific frequency bands, covered in detail in section 6.3, corresponding to different physical areas within the brain. As discussed in Chapter 4 the signals are stochastic, i.e. they do not repeat themselves as a function of time: indeed at first sight the EEG signal can appear very similar to random noise. A complete EEG recording system consists of scalp electrodes connected to units for amplification/filtering, analogue-to-digital conversion, signal processing, and real-time data display and storage. Since the SNR of the EEG signal is relatively low, a number of mathematical algorithms have been developed to try to tease out subtle characteristics of the signals produced by specific clinical conditions. For example, the individual signals from each channel can be denoised, and then cross-correlated with other channels to study the interactions of different areas of the brain. Wavelet and other transforms can be applied to try to determine the most important components of the signals. More details on these types of algorithms can be found in texts such as Sanei and Chambers [1].

Clinically, an EEG is used to assess the general health of the higher central nervous system (CNS). The major applications are:

(i) epilepsy detection and monitoring
(ii) monitoring the depth of anaesthesia during surgery in the intensive care unit using the bispectral index
(iii) monitoring brain activity during surgery in traumatic brain injury, and
(iv) assessing brain health with respect to hypoxic–ischaemic brain injury after a cardiac arrest or post carotid artery surgery.

In addition to these direct clinical applications, EEG signals are increasingly being used in therapeutic devices, which come under the broad title of brain–computer interfaces (BCI). Signals from the brain such as visual evoked potentials and sensorimotor rhythms can be used to control prosthetic limbs in severely injured patients. This topic is covered in the final section of this chapter.

6.1 Electrical Signals Generated in the Brain

The topic of action potentials and their propagation through tissue was first discussed in section 2.2.2. In the following sections the location of areas of the brain that produce action potentials is outlined, and the implications of homogeneous and inhomogeneous volume conductors discussed.

6.1.1 Postsynaptic Potentials

The outer section of the cerebrum is termed the cerebral cortex, which forms a highly folded surface with a thickness between approximately 2 and 5 mm, as shown in Figure 6.2(a). The cerebral cortex is composed of grey matter (so-called because of its appearance in visible light during dissection). The cerebral cortex generates the majority of the electrical potentials that are measured by EEG electrodes. It contains more than 10 billion neurons, which have a high degree of connectivity. A large cortical neuron may have between 10^4 and 10^5 connecting synapses that transmit signals from other neurons. These synapses can be of two different types, one which produces an excitatory postsynaptic potential (EPSP) across the membrane making it easier for an action potential to fire, and the other which produces the opposite effect, namely an inhibitory postsynaptic potential (IPSP).

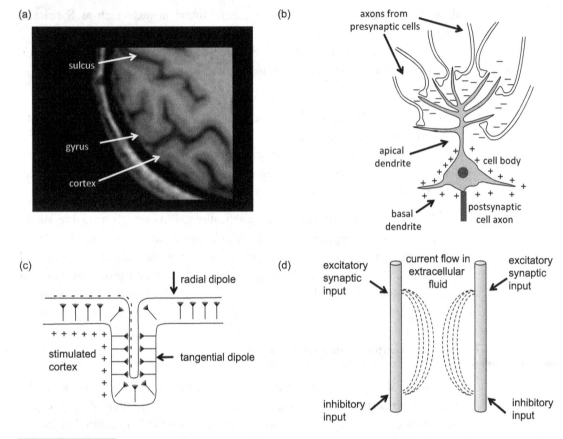

(a)

sulcus

gyrus

cortex

(b)

axons from
presynaptic cells

apical
dendrite

cell body

basal
dendrite

postsynaptic
cell axon

(c)

radial dipole

stimulated
cortex

tangential dipole

(d)

excitatory
synaptic
input

current flow in
extracellular
fluid

excitatory
synaptic
input

inhibitory
input

inhibitory
input

Figure 6.2 (a) MRI of the brain showing the cerebral cortex, gyri and sulci. (b) Schematic of a pyramidal cell and dendrite. The cell body has dimensions of approximately 20 μm. The length of a single dendrite is usually several hundred μm. The total length of the dendrite and axon of a pyramidal cell may reach several centimetres. (c) Pyramidal cells in the gyri act as radial dipoles, whereas those in the sulci act as tangential dipoles. The EEG signal is generated mainly by the cells in the gyri. (d) Membrane potential differences between the excitatory and inhibitory synaptic inputs generate current flow in the extracellular fluid.

Within the cerebral cortex, the largest component of the EEG signal is generated by pyramidal cells (also called pyramidal neurons). Pyramidal neurons receive synaptic inputs from tens of thousands of excitatory synapses and several thousand inhibitory synapses. Most of the excitatory inputs use glutamate as the neurotransmitter, while inhibitory inputs use gamma-amino butyric acid (GABA). As shown in Figure 6.2(b) the pyramidal cell has a long straight apical dendrite (the pyramidal cell dendrite), which extends up through the cortical layers from the cell body directly towards the pial surface of its gyrus. Neighbouring pyramidal cells contain dendrites that are roughly parallel to one another. Current dipoles are created in the gyri and sulci, as shown in Figure 6.2(c). The physical separation of inhibitory and excitatory PSPs on an apical dendrite creates bridging current loops between the PSPs. When neighbouring pyramidal cells have similar and synchronous areas of altered membrane potentials, their current loops combine together in the extracellular fluid to create a much larger regional current flow, as shown in Figure 6.2(d).

6.1.2   Volume Conduction Through the Brain

The postsynaptic current described in the previous section propagates through the cerebrospinal fluid (CSF), the skull and the scalp, where it is detected by many electrodes located on the scalp surface, with the exact positions referenced to the anatomy of the particular patient (covered in section 6.2.1). The electrodes can either be unipolar or bipolar. Unipolar electrodes reference the potential measured by each electrode to either the average of all the electrodes or to a 'neutral' electrode. Bipolar electrodes measure the difference in voltage between a pair of electrodes.

The CSF and scalp are relatively conductive compared to the skull, and the overall effect of transmission through these layers is a substantial spatial smearing of regional voltage differences. This means that the signal from an EEG electrode reflects activity over a wide area, not just directly under the electrode. The relationship between the potentials measured by the scalp electrodes and the underlying sources of current in the brain generated in a homogeneous conducting medium can be expressed by the following form of Poisson's equation:

$$\nabla \bullet [\sigma(r)\nabla\Phi(r,t)] = -s(r,t) \tag{6.1}$$

where $\nabla \bullet$ represents the divergence operator and ∇ the gradient operator, $\sigma(r)$ is the tissue electrical conductivity and $s(r,t)$ is the current source function in the neural tissue, as shown in Figure 6.3.

Rather than dealing with the actual microsources of the signals, an intermediate scale of sources is defined as $P(r',t)$, corresponding to the signal in 'mesoscopic' cortical columns, making use of the known cortical electrophysiology to relate P to the original brain microsources. The quantity P is termed the current dipole moment per unit tissue volume, and has units of current density (microamps per square millimetre).

$$P'(r,t) = \frac{1}{W}\int\int\int_{w} ws(r,w,t)dW(w) \tag{6.2}$$

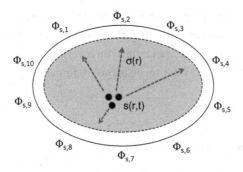

Figure 6.3 Simple schematic of the propagation of ionic currents created by pyramidal cells, as shown in Figure 6.2, through a homogeneous volume conductor with conductivity σ to the scalp, where they are detected by surface electrodes as surface potentials Φ_s. The filled circles represent regions where current sources s(r,t) are present. Figure adapted from Nunez [2].

where the transformation of s(r,t) to s(r,w,t) indicates an integration over a tissue volume W with its centre located at r. For example, if W represents a cortical column then P(r′,t) is the diffuse current density across this column. Approximately 1000 macrocolumns (containing approximately 1 million neurons each) are required to generate a sufficient dipole moment P(r′,t) to be detectable as a scalp potential. Typically, this corresponds to about 5 square centimetres of cortical gyrus tissue, although the exact value depends upon the distance to the skull and also the degree of synchronization within the macrocolumn assembly.

Surface potentials are defined as:

$$\Phi_s(r,t) = \iint_S G_s\left(r,r'\right) \cdot P\left(r',t\right) dS\left(r'\right)$$

(6.3)

where G represents the appropriate Greens function. The recorded potential, V, between a pair of scalp electrodes with positions r_i and r_j, respectively, can be represented as:

$$V\left(r_i, r_j, t\right) = \Phi(r_i, t) - \Phi(r_j, t)$$

(6.4)

In fact, most potentials are measured with respect to a reference location r_R on the head or neck:

$$V(r_i, r_R, t) = \Phi(r_i, t) - \Phi(r_R, t)$$

(6.5)

As mentioned previously, the net effect of volume conduction is that the potential at the scalp is 'smeared' significantly, and so several electrodes may be sensitive to a particular event that happens at one location in the brain. This is

Figure 6.4

(a)

(b)

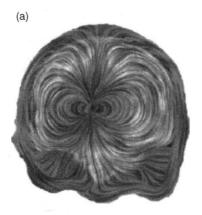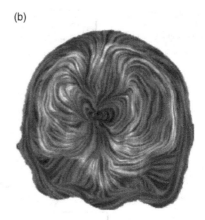

Surface potentials produced by a single event in the thalamus. Two results are shown: (a) with a completely isotropic brain used as the volume conductor, and (b) second with a more realistic model with white matter specifically modelled.

illustrated in Figure 6.4, which shows the large spatial extent of surface potentials from a single event in the thalamus. The differences between Figures 6.4(a) and (b) also shows the effects of assuming an isotropic homogeneous brain model, or a more realistic one that incorporates the known anisotropy of the properties of white matter.

As PSPs occur and decay, the EEG scalp voltages change over time. Millions of PSPs occur all over the cortex, producing a continuously time-varying EEG. As a result, unlike the ECG signal, the normal EEG signal has no obvious repetitive patterns, nor does the shape of the EEG waveform correlate with specific underlying events. Indeed, higher cortical function is usually associated with **desynchronization**, as neurons act more independently when 'creating' conscious human behaviour as opposed to being at rest. Indeed, anaesthesia and other mechanisms that depress consciousness are associated with increasing **cortical synchrony**. However, under some conditions the EEG may contain characteristic waveforms that can be used diagnostically. For example, 'spikes' or 'sharp waves' are created by a high degree of transient synchrony, and these features can be used in the diagnosis of epilepsy, covered in section 6.4.1.

6.2 EEG System Design

The overall design of an EEG system is shown in Figure 6.5. The design is quite similar to that of an ECG system covered in Chapter 5, the main differences being that many more channels are used, and that the signals have lower amplitudes and dynamic range, thus requiring a higher degree of amplification and filtering.

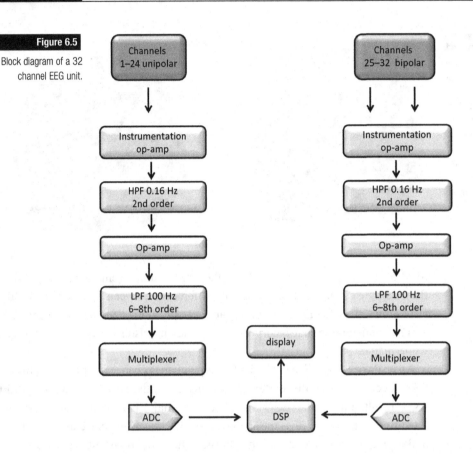

Figure 6.5

Block diagram of a 32
channel EEG unit.

6.2.1   Electrodes and their Placement on the Scalp

The most common arrangement of scalp electrodes is the International 10–20
system (Figure 6.6). The numbers 10 and 20 indicate that the distances between
adjacent electrodes are either 10% or 20% of the total front–back or right–left
dimensions of the skull. The coordinate axes for the 10–20 systems are based on
meridians crossing the scalp. Two anatomical landmarks are used for positioning
the EEG electrodes: the nasion, which is the depressed area between the eyes just
above the bridge of the nose, and the inion, which is the lowest point of the skull at
the back of the head and is normally indicated by a prominent bump.

Electrodes are typically a Ag/AgCl design, very similar to those used for ECG,
although smaller in size (diameter 4 to 10 mm). Conductive gel containing
a concentrated salt solution is applied to the skin. Scalp electrode impedance is
reduced by abrading the scalp under each electrode prior to electrode placement.
If only a few electrodes are used, the skin under each electrode can be abraded
using cotton swabs dipped in alcohol or abrasive paste. If a large number of

(a)

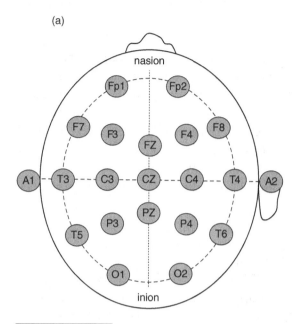

(b)

Figure 6.6   (a) Schematic showing the positioning of the International 10–20 system of electrode placement. F refers to frontal lobe, T to temporal lobe, C to central lobe, P to parietal lobe and O to occipital lobe. A (aural) represents the ear. Electrodes on the right side of the brain have even numbers, and those on the left have odd numbers. The numerals indicate the relative distance from the midline. The letter Z indicates an electrode placed on the midline. (b) Photograph of an EEG cap containing a full set of electrodes, which are often sewn into the cap.

electrodes are used (i.e. 32 or more), the scalp electrodes are typically attached to an elastic cap or net. Once the cap is in place, the skin under each electrode is abraded using a blunt needle or wooden stick that is inserted through a hole at the back of each electrode. According to the International Federation of Clinical Neurophysiology guidelines and the American Clinical Neurophysiology Society recommendations, the impedance of each electrode should be checked before every recording and should ideally be between 1 kΩ and 5 kΩ. If the impedance is too high, then the noise in the recordings is increased.

Other types of electrode can be used, including dry electrodes [3], but these are currently considered to be less robust. Active electrodes, which have the amplifier directly at the electrode, can also be used: these still require conducting gel, but skin abrasion is not so critical. In the ICU, cup electrodes are usually glued onto the scalp. This type of electrode is highly flexible and can be placed on patients with skull defects, or with intracranial pressure monitoring devices. These electrodes can be used for standard short timescale EEG recordings (for tens of minutes) but also provide stable impedances for prolonged continuous EEG recordings, although after ten days a period of two to three days without electrodes is recommended to avoid skin damage. Plastic cup electrodes, which can remain attached if the patient has to undergo a computed tomography or magnetic resonance imaging scan, are also commercially available. Long time-frame continuous EEG (cEEG) may be required in patients who, for example, only show very infrequent episodes of epileptic events, which are therefore easily missed by short timescale recordings. Typical periods for cEEG might be 24 hours for non-comatose patients and 48 hours for comatose patients. For cEEG, a very robust contact is needed. Collodion, which is a solution of nitrocellulose in ether and alcohol, is the most commonly used adhesive. When applied onto the skin, the alcohol and the ether evaporate, forming a thin nitrocellulose film, which attaches the electrode to the scalp.

6.2.2 Amplifiers/Filters and Digitizing Circuitry

Signals from the EEG electrodes have an amplitude of the order of tens of microvolts spanning a frequency range of DC to 50 Hz, as shown in Figure 6.1. In addition to the EEG signal, there are three main interfering physiological signals, which can be picked up as potential differences between two EEG electrodes.

(i)   an ECG signal corresponding to the R-wave travelling through the neck
(ii) an EMG signal from the electromechanical activity of muscle in the scalp, which has frequency components up to 100 Hz, and
(iii) an EOG signal generated as the eyes move.

In addition, there is the usual 50/60 Hz noise contribution from radiating power lines or cables close to the system.

Figure 6.7 is a schematic showing the combined amplification and filtering circuitry used for an EEG acquisition system. An instrumentation op-amp, covered in section 3.3.3, reduces common-mode signals. The input impedance of this amplifier should be very large compared to that of the scalp–electrode interface to limit the current flowing through the system (via a high-impedance mismatch). Limiting the current minimizes the voltage drop across the scalp–electrode interface, i.e. minimizes the SNR loss. Modern EEG amplifiers have input impedances that have a real part of the order of 200 MΩ. Assuming that the scalp–electrode interface has an impedance of 5 kΩ, this means that the maximum signal loss is 0.025%.

An in-line HPF and LPF are also shown in Figure 6.7. A Sallen–Key architecture is used to minimize the number of components needed for a Butterworth-type low-pass filter. The total gain of the receiver chain is set to a factor of approximately 1000.

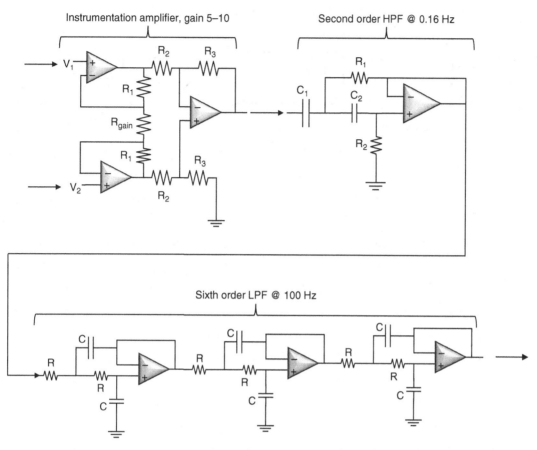

Instrumentation amplifier, gain 5–10

Second order HPF @ 0.16 Hz

Sixth order LPF @ 100 Hz

Schematic of the different subunits for the amplification and filtering of the EEG signal before digitization.

Most digital EEG systems sample at rates between 240 and 512 Hz to avoid aliasing of any EMG signals that are not completely filtered out. In most EEG systems time-domain multiplexing is used, in other words the signals from multiple channels are recorded using a single ADC, with the signal inputs being switched between channels by a MUX, as shown in Figure 6.8. This approach reduces cost and increases the simplicity of the circuitry, but one must take into account the fact that different channels are sampled at slightly different times. Using a sampling rate of 250 Hz, for example, the signals from different channels would have a 40 ms delay (sampling skew) between each one. This effect can be minimized by using what is termed **burst mode sampling**, which is also illustrated in Figure 6.8. The sampling skew is now dictated by the much shorter Δt_{burst} rather than the longer $\Delta t_{channel}$. Typical processing and switching times for the individual components shown in Figure 6.8 are less than 100 ns for the analogue multiplexer,

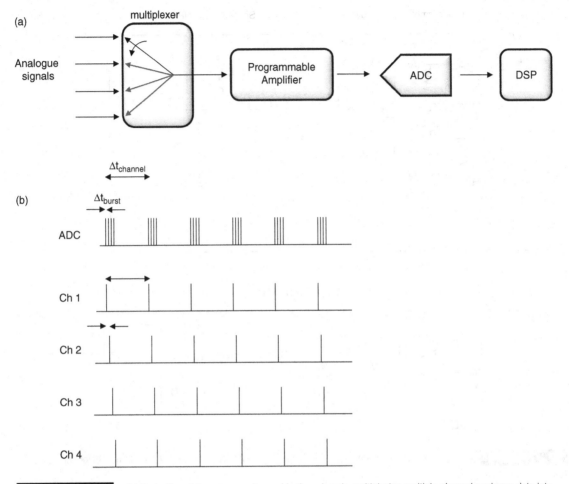

Figure 6.8 (a) Illustration of the components used in time-domain multiplexing multiple-channel analogue data into a single ADC. (b) An example of burst mode sampling, in which the ADC samples at a high rate for a short period of time. The sampling time per channel is $\Delta t_{channel}$, but the inter-channel sampling skew is given by Δt_{burst}, which is much less than $\Delta t_{channel}$.

~200 ns for the programmable amplifier and a few microseconds for the ADC. This means that the ADC can operate at a burst frequency of several hundreds of kHz.

6.3 Features of a Normal EEG: Delta, Theta, Alpha and Beta Waves

EEG signals can be classified into different frequency bands, each of which have specific physiological properties or have different spatial distributions within the brain. Illustrative examples are shown in Figure 6.9.

Figure 6.9

Schematic examples of EEG waveforms in different frequency bands.

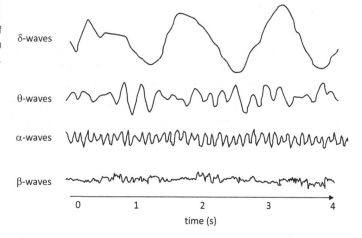

(i) The delta (δ) band consists of frequencies below 4 Hz. Delta rhythms are usually only observed in infants or in adults who are in a deep sleep state. A large amount of delta activity in awake adults is abnormal and is often associated with neurological diseases. Delta waves have their sources either in the thalamus or in the cortex.

(ii) Theta (θ) waves are defined to lie in the 4 to 7 Hz range. In a normal awake adult, only a small amount of theta waves occur. A larger amount of theta waves is seen in young children, older children and adults in light sleep states. As is the case for delta waves, a large amount of theta activity in awake adults is related to neurological disease. The theta band is also associated with a wide range of cognitive processes.

(iii) Alpha (α) waves have frequencies between 8 and 12 Hz and can be measured over the occipital region in the brain. Their amplitude increases when the eyes close and the body relaxes, and attenuates when the eyes open and mental tasks are being performed. Alpha waves primarily reflect visual processing in the occipital brain region and may also be related to memory recollection.

(iv) Beta (β) waves lie between 12 and 30 Hz and can be recorded from the frontal and central regions of the brain. Beta waves are characterized by their symmetrical distribution in the waveform peaks when there is no motor activity, and become desynchronized (asymmetric) and have a lower amplitude during movement or motor 'imagery'.

Other types of waves such as gamma (γ) waves, which are found in the frequency range from 30 to 100 Hz, have been used for neuroscience investigations, but are not routinely assessed in clinical practice.

6.4 Clinical Applications of EEG

The two major clinical applications of EEG are in epilepsy diagnosis and treatment monitoring, and for monitoring anaesthesia levels during surgery, each of which is detailed in the following section. There are many other applications, including the management of comatose patients with subarachnoid and intracellular haemorrhage, or traumatic brain injury. EEGs are also often acquired from patients who have unexplained changes in conscious states (e.g. sudden blackouts).

6.4.1 EEG in Epilepsy

One of the major clinical applications of EEG is the diagnosis, monitoring and clinical treatment of epilepsy [4]. Epilepsy is the most common neurological disorder, affecting 50 million people worldwide, 85% of which live in developing countries. Around 2.4 million new epileptic cases occur every year globally and at least 50% begin at childhood or adolescence. Epilepsy is a chronic brain disorder, characterized by seizures, that can affect any person at any age. Epileptic episodes may occur with a frequency as low as once a year up to several times per day. In general, epileptic episodes are highly unpredictable, and for more than 60% of cases no definitive cause can be ascertained. While not preventable, epilepsy is treatable with antiepileptic medications. Clinical diagnosis of epilepsy requires a detailed case history, neurological examinations as well as blood and other biochemical tests. Epilepsy is characterized by unprovoked seizures due to the process of 'epileptogenesis' in which neuronal networks abruptly turn hyperexcitable, affecting mostly the cerebral cortex. This leads to motor function abnormalities, causing muscle and joint spasms.

EEG signals recorded just before and during epileptic events contain patterns that are different from those in a normal EEG signal. EEG analysis can not only differentiate epileptic from normal data, but also distinguish different stages of an epileptic episode, such as pre-ictal (EEG changes preceding an episode) and ictal (EEG changes during an episode): an example of the latter is shown in Figure 6.10. In cases where patients have more than one seizure within a short period of time, there is an additional stage called the inter-ictal stage, also shown in Figure 6.10. During the pre-ictal stage, there is a reduction in the connectivity of neurons in the epileptogenic zone. The epileptic neurons become isolated from the circuit, become idle and lose inhibitory control. This results in an epileptic episode caused by a sudden increase in neural discharge.

There is growing interest in computer-aided diagnosis (CAD) of epilepsy using EEG signals. Different time- or frequency-domain techniques can be used to

(a)

(b)

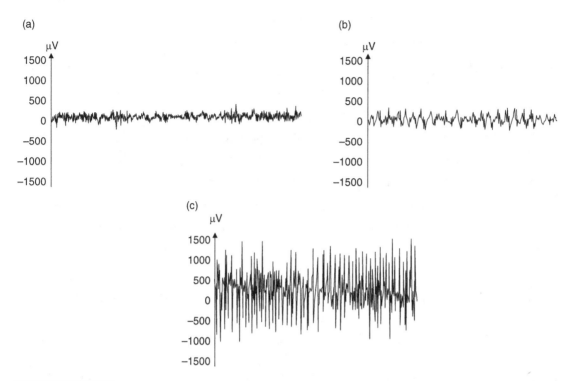

(c)

Figure 6.10 Typical EEG signals (a) normal, (b) inter-ictal and (c) ictal.

analyze long periods of continuous EEG recordings. Techniques such as principle and independent component analysis (PCA and ICA, respectively), and linear discriminant analysis can be used to distinguish features in the EEG trace that are linked to epilepsy [5]. In terms of possible surgical resection of parts of the brain where epileptic episodes are centred, the topic of localization of the sites of epileptic events using electric source imaging is covered in section 6.6.

6.4.2 Role of EEG in Anaesthesia: the Bispectral Index

The second main clinical application of EEG is to monitor the depth of anaesthesia in the operating theatre or ICU. Examples of EEG traces reflecting various degrees of anaesthesia are shown in Figure 6.11.

An EEG acquired during anaesthesia typically results in a relative increase in the delta-wave component and a decrease in the beta-wave component compared to when the patient is awake, as shown in Figure 6.11(a)–(d). One can quantify this by measuring the SEF95 (the frequency below which 95% of the signal occurs),

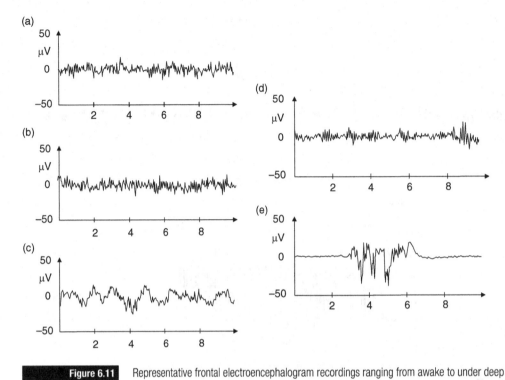

Figure 6.11 Representative frontal electroencephalogram recordings ranging from awake to under deep anaesthesia. (a) Awake, (b) light anaesthesia, (c) surgical anaesthesia – delta spindles are present, (d) surgical anaesthesia – showing the presence of sleep spindles, (e) surgical anaesthesia – showing periods of burst suppression. In general, higher frequencies predominate during the awake state, while slower delta waves become more prominent in surgical anaesthesia. Sleep spindles, or short bursts of alpha waves, are also commonly seen during general anaesthesia. Burst suppression, or intermittent bursts of electrical activity amid periods of persistent suppression, occurs during very deep anaesthesia.

which shifts from above 20 Hz to below 12 Hz. During deep anaesthesia, a phenomenon known as burst suppression may also occur, shown in Figure 6.11(e). This phenomenon appears as successive periods of normal, or higher than normal, amplitude EEG signals alternating with periods of very low signals close to the noise floor. The burst suppression ratio (BSR) is defined as the fraction of time for which burst suppression is present, where a period of burst suppression is defined as any period longer than 0.5 seconds during which the magnitude of the EEG voltage is below 5 µV. Since the EEG signal is non-stationary (section 4.6.1), the BSR is defined over a time period of at least one minute. A composite measure, the burst-compensated spectral edge frequency (BcSEF), is defined as:

$$BcSEF = SEF\left(1 - \frac{BSR}{100}\right) \tag{6.6}$$

Figure 6.12

Schematic showing the
various steps in
determining the bispectral
index.

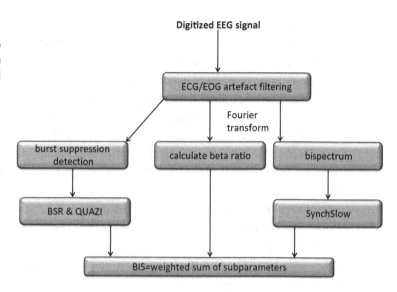

Ideally there would be a quantitative measure of the depth of anaesthesia during surgery, which could be monitored in real time throughout the procedure. Building upon concepts such as the BcSEF, the bispectral index (BIS) [6–8] is currently the only quantitative measure that is approved by the FDA for monitoring the effect of anaesthesia on the brain [9]. It represents a combination of a number of different measures, including time-domain, frequency-domain and high-order spectral sub-parameters. The calculation of BIS is shown schematically in Figure 6.12.

The digitized EEG signal is first filtered to exclude artefacts, and divided into epochs of 2 seconds' duration. The first component of artefact rejection uses cross-correlation of the EEG epoch with a template pattern of an ECG waveform. If an ECG signal is detected, it is removed from the epoch, and the missing data is estimated by interpolation. Epochs repaired in this way are still considered viable for further processing. In the next step, eyeblink events are detected, based on a similar template-matched cross-correlation technique, this time using an EOG template. Since this type of interference produces a high level of signal distortion, epochs with blink artefacts are not processed further. After these two signal-processing steps have been performed the epochs are checked for a varying baseline (due to low-frequency electrode noise), and if present then additional high-pass filtering is applied to remove this effect. The variance of the EEG waveform for each epoch is then calculated. If the variance of a new epoch is significantly different from an average of the variances of recently recorded epochs, the new epoch is marked as 'noisy' and not processed further; however, its variance is incorporated into an updated average.

Assuming that the EEG epoch is artefact free or interpolation has been performed to fill-in missing data, three separate 'branches' of data processing are performed, as shown in Figure 6.12. In the left-hand branch the time-domain signal is used to calculate the degree of burst suppression with two separate algorithms: BSR and 'QUAZI'. The BSR algorithm used by the BIS calculation is similar to that described in the previous section. The QUAZI suppression index is designed to detect burst suppression in the presence of a slowly varying baseline voltage. QUAZI works by high-pass filtering to remove the baseline, which would otherwise 'fool' the original BSR algorithm by exceeding the voltage criteria for electrical 'silence'.

In the central and right branches of Figure 6.12 data is multiplied by a window function, and then the FFT and the bispectrum of the current EEG epoch are calculated (the bispectrum is mathematically defined as a third-order higher statistic). The resulting spectrum and bispectrum are smoothed using a running average against those calculated in the previous minute. Finally, the frequency domain-based subparameters 'SynchFastSlow' and 'Beta Ratio' are computed. The Beta Ratio sub-parameter is defined as the log of the power ratio in two empirically derived frequency bands (30 to 47 Hz and 11 to 20 Hz):

$$\beta_{ratio} = \log\left(\frac{P_{30-47Hz}}{P_{11-20Hz}}\right) \tag{6.7}$$

The SynchFastSlow sub-parameter is defined as another log ratio, this time the sum of all bispectrum peaks in the range from 0.5 to 47 Hz divided by the sum of the bispectrum in the range 40 to 47 Hz.

$$synchfastflow = \log\left(\frac{P_{0.5-47Hz}}{P_{40-47Hz}}\right) \tag{6.8}$$

The combination of the four sub-parameters BSR/QUAZI/SynchFastSlow and Beta Ratio produces a single number, the BIS. These algorithms effectively produce different adaptive weightings for the four parameters. The SynchFastSlow parameter is well correlated with behavioural responses during moderate sedation or light anaesthesia and is therefore weighted most heavily when the EEG has the characteristics of light sedation. In contrast, the BSR and QUAZI parameters are more effective at detecting deep anaesthesia. The reported BIS value represents an average value derived from the previous 60 seconds of useable data. The BIS scale runs from 90 to 100, which corresponds to awake, to less than 40, which represents deep anaesthesia. The overall strategy in the operating theatre and ICU is typically to aim to maintain the BIS between values of 45 and 65.

6.5   EEG Signals in Brain–Computer Interfaces for Physically Challenged Patients

One of the most recent, and still very much in development, medical applications of EEG is to brain–computer interfaces (BCIs) [10–14], also called brain–machine interfaces (BMIs), which are systems that allow a severely physically challenged patient to control an external device based upon brain activity [10, 11]. The electrical signals of the brain, sensed via an EEG system, are used as the input to such an interface. The major challenge is how to interpret the EEG signals. Ideally, non-invasive EEG measurements would be used, since they do not require electrode implantation with the associated problems of rejection or infection. However, implanted electrodes do have intrinsic advantages in terms of a much higher SNR and much higher degree of localization. In order to use EEG signals to control a computer interface, one can break down the requirements into five steps:

(i) signal acquisition
(ii) signal preprocessing and filtering
(iii) feature extraction
(iv) feature classification, i.e. the process by which the patient's intention is recognized on the basis of one or more of the features extracted from the EEG data, and
(v) interface control.

The first two steps, signal acquisition and signal preprocessing and filtering, are essentially identical to those in standard EEG, using the same hardware. The signals, shown in Figure 6.13, that are currently used as inputs to control algorithms in BCIs are visual evoked potentials (VEPs), slow cortical potentials (SCPs), P300 evoked potentials and sensorimotor (mu and beta) rhythms.

(i) VEPs are changes in the EEG signal that occur in the visual cortex upon reception of a visual stimulation. The magnitude of the VEP depends upon the position of the stimulus within the visual field, with a maximum response corresponding to the centre of the visual field. Steady-state VEPs (SSVEPs) occur when the stimuli are presented at a rate higher than about 6 times per second.
(ii) SCPs are very low-frequency (<1 Hz) changes in the EEG signals, with a negative SCP being produced by increased neuronal activity, and vice versa.
(iii) P300 evoked potentials are positive voltage peaks in an EEG trace that occur approximately 300 ms after attending to a particular targeted stimulus, which is presented among other random stimuli.
(iv) Signals in the mu band (7 to 13 Hz) and beta band (13 to 30 Hz) are related to motor imagery, i.e. imagined motion without any actual movement.

Figure 6.13

Illustration of four different measured EEG signals that can be used as inputs to a brain–computer interface: (a) the P300 evoked potential, (b) the visual evoked potential, (c) slow cortical potentials, and (d) beta/mu frequency bands corresponding to sensorimotor rhythms.

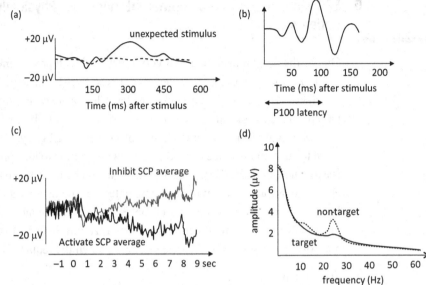

Brain–computer interfaces can be characterized as either exogenous, i.e. requiring an external stimulus to evoke neuronal activity, or endogenous, in which the stimulus is generated internally. Steady-state VEPs and P300 signals are exogenous, whereas SCPs and sensorimotor rhythms are endogenous. Applications of each of these signals are covered in the next section.

6.5.1   Applications of BCIs to Communication Devices

Many of the applications of EEG-based BCI are in communications, i.e. enabling a patient with severe physical challenges to 'type' sentences. This process is carried out using a virtual keyboard on a screen where the patient chooses a letter using a BCI. There are many methods of doing this, the most common being the P300 speller and variations thereof. P300 event-related brain potentials can be used by the patient looking at a 6 × 6 matrix of letters that flash at random during an EEG measurement. The patient counts how many times the row or column containing the intended letter flashes. The P300 signal is detected only when the target letter is in the displayed column or row. The raw EEG signals are filtered with a band-pass digital filter with cut-off frequencies of 0.1 Hz and 30 Hz and averaged together to improve the SNR. Feature analysis is then performed, using a variety of

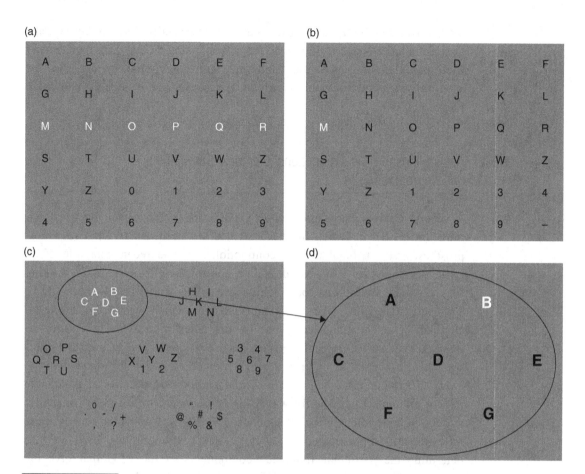

Figure 6.14   Different variations of the P300 speller. (a) Row/column paradigm. (b) Single-character paradigm. (c, d) Region-based paradigm: (c) first step and (d) second step in the region-based paradigm.

different methods including wavelet transforms, ICA and PCA. Finally, the waveforms are classified into whether they contain a P300 or not. The method is quite slow, since it requires multiple repetitions per letter. Many variations and improvements on this basic theme exist, including optimized groupings of letters and numbers: examples are shown in Figure 6.14.

Voluntary control of SCPs can also be used for letter selection. After extensive training, completely paralyzed patients are able to either produce positive or negative changes in their SCPs, as shown in Figure 6.13(c), to drive the vertical movement of a cursor. A thought-translation device is used for training the patient to control their SCP signals, although this process typically takes several months. Two characters per minute is about the limit of this approach. Steady-state VEPs can also be used by patients to select a target, such as letters on the screen, using an eye-gaze mechanism. EEG modulation via mental hand and leg motor imagery can also be used for each subdivision, based on the sensorimotor signals shown in Figure 6.13(d).

6.5.2 Applications of BCIs in Functional Electrical Stimulation and Neuroprostheses

A second clinical application of EEG–BCIs is to functional electrical stimulation (FES), in which the EEG signals are used as the input to neuroprostheses via artificial muscle contractions. In one recent example, King *et al.* developed an EEG-based BCI to control an FES system for walking for a patient who was paraplegic due to a spinal cord injury [15]. The particular FES used is called Parastep, which walks via electrical stimulation of the quadriceps and tibialis anterior muscles to control an exoskeleton. An EEG-decoding model must be developed in order to interpret the EEG data to control the FES device in order for the patient to 'walk'. Training takes place in a virtual-reality environment in which the patient attempts to 'mentally walk' for 30 seconds, followed by 'mentally sitting' in a wheelchair for 30 seconds. EEG data is collected during these intervals and sophisticated methods of data extraction are used [15] to determine the particular frequency components of the EEG that give the greatest difference between 'walking' and 'sitting', and so can be used to provide an input to the neuroprostheses. Some of the spatial features produced during this process are shown in Figure 6.15, with discriminating features between walking and sitting seen under electrodes CP3, CPz and CP4.

Although the work by King *et al.* represents a very early study, the results are highly promising with the patient achieving a high level of control and maintenance over a 19-week period. Other researchers have shown how EEG recordings can be used to restore some ambulatory control to another patient with spinal cord injury [15, 16]. Limited control of devices such as wheelchairs can also be induced by using P300s, although the latency due to the required extensive averaging of low SNR signals can be a problem. Neurofeedback based on EEG signals has also been used in conditions such as pain management, and mental disorders such as depression and schizophrenia.

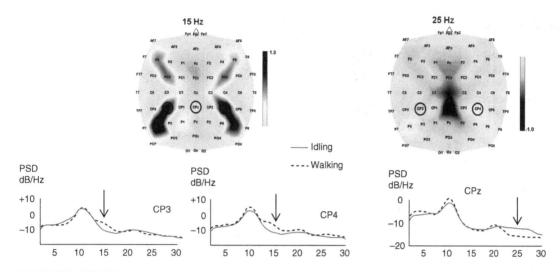

Figure 6.15 EEG feature-extraction maps produced during walking and sitting. The spatial distribution of features is shown for the frequency bands centred at 15 Hz and 25 Hz, where the features with values close to ±1 are more important for classification. Log power spectral density (PSD) plots are shown during idling and walking at electrodes CP3, CPz and CP4. The arrows show the event-related synchronization (ERS) in the 13 to 16 Hz range (at CP3 and CP4) and event-related desynchronization (ERD) in the 23 to 28 Hz range (at CPz). Figure reproduced with permission from King *et al.* [15].

6.6   Source Localization in EEG Measurements (Electrical Source Imaging)

In patients with focal epilepsy who do not react to drug therapies, a condition referred to as pharmacoresistant focal epilepsy, surgical resection of the epilepto-genic volume is the medical procedure of choice. Precise pre-surgical localization of the epileptogenic volume is a key element in removing as many of the foci, while sparing as much of the surrounding healthy brain tissue, as possible. Electrical source imaging (ESI) is a technique by which EEG signals recorded on the scalp, as covered in this chapter, are co-processed with other imaging data such as MRI scans in order to try to localize the sources of the inter-ictal spikes, i.e. the epileptic foci [17–20].

As covered in section 6.1.2 on volume conduction, the relationship between the EEG signals measured at the scalp and the electrical source within the brain depends upon the geometry and heterogeneity of the volume conductor, i.e. the brain, CSF and scalp. One of the key steps in ESI is to obtain a realistic model of the volume conductor, and this can be achieved using an MRI of the patient, which is obtained as part of the standard diagnostic procedure. The MRI can then be segmented into areas of bone, brain tissue (white and grey matter) and CSF to produce the volume conductor model.

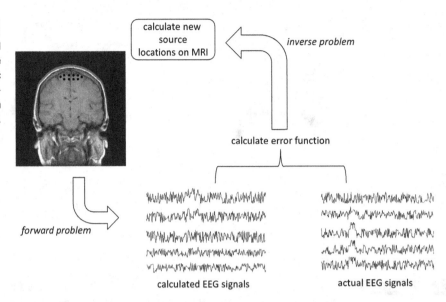

Figure 6.16

Schematic of the ESI method used to derive the localized foci of epileptic seizures from the inter-ictal spikes recorded on the EEG.

The EEG acquisition system used for ESI is very similar to that outlined in this chapter, except that typically a much denser array (128 or 256) of smaller electrodes is used in order to increase the reconstruction accuracy. The EEG signal is recorded continuously for approximately half-an-hour, and then a neurologist will determine which electrode signals show inter-ictal spikes.

The first computational step in ESI is called the **forward problem**, which models how the electrical currents propagate through the head, i.e. given a certain current distribution what would the EEG signals be after propagation through the head model obtained from the patient's MRI. Current sources are constrained to lie in the grey matter, and are initially placed on a regular grid throughout the brain, approximately every 5 mm, as shown in Figure 6.16.

Having determined the theoretical EEG signals that would have been produced from the initial electric current sources within the brain, the next step is to compare the actual EEGs with the theoretical EEGs. An error signal (the difference between the modelled and measured data) is calculated, and then the **inverse problem** starts to be solved, i.e. given the actual EEGs what is the actual distribution of current sources that produced this distribution? Mathematically, this corresponds to minimizing the error signal. There are many different algorithms that can be used to perform this minimization, each with its own advantages and disadvantages [21]. The process is mathematically highly intensive, and involves many hundreds or thousands of iterations, and so requires significant computing power. Since there is no unique solution physiological bounds, for example the maximum possible EEG signal strength or relative co-location of EEG signals, can be integrated into the solution. The accuracy with which the epileptogenic source can be located is affected by a number of factors including any head-modelling errors or noise in the EEG signals [22]. Despite the complexity of the problem a number of demonstrations have been shown in clinical cases, with successful resection of the epileptogenic sources.

PROBLEMS

6.1 Classify the two EEG traces shown in Figure 6.1(a) and (b) as either deterministic or stochastic. If stochastic, further classify as either stationary or non-stationary.

6.2 Explain why the EEG signal is generated mainly by cells in the gyri rather than those in the sulci.

6.3 Supposing that there is an area of high conductivity (shown as the dark-shaded area in the figure below) within the brain. How would this affect the potentials recorded by the EEG electrodes around the head?

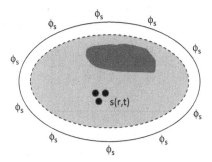

6.4 Assume that on a clinical EEG the input impedance for the amplifiers is 10 MΩ. Calculate the SNR loss if the scalp–electrode impedance is 10 kΩ, 100 kΩ and 1 MΩ.

6.5 Some EEG systems use active electrodes, which integrate high-input impedance amplifiers into the electrode. Describe the potential advantages and disadvantages of such an approach.

6.6 Calculate values for the capacitors and resistors for the Sallen–Key filters shown in Figure 6.7.

6.7 For a 32-channel EEG system with 10 μs sampling skew, what is the minimum sampling rate required for the ADC?

6.8 As for Problem 6.1, classify the five EEG traces shown in Figure 6.11 as either deterministic or stochastic. If stochastic, further classify as either stationary or non-stationary.

6.9 By reading reference [5] by Acharya *et al.*, choose one technique for automatic analysis of the EEG signal for detecting epilepsy. Explain the basic method and the experimental requirements. Outline the advantages and disadvantages of the particular technique.

6.10 Suggest a simple method by which a long continuous EEG signal could be analyzed to detect if burst suppression had occurred? What would be the instrumental requirements to make your method stable?

6.11 Investigate what type of template can be used to remove EOG signals from the EEG signal.

6.12 Explain how each of the BCI-based communication systems shown in Figure 6.14 work.

6.13 Suggest a new design for a BCI-based communication device based either on the P300 signal or slow cortical potentials. Indicate what type of instrumentation would be necessary.

6.14 Returning to the situation in Problem 6.3, what are the effects of the (unknown) increase in conductivity of part of the brain on a general electric source imaging algorithm?

6.15 Name three constraints that could be used to better characterize the solution to the ESI problem.

6.16 EEG data is increasingly being recorded simultaneously with MRI data. By looking in the literature, determine what challenges this involves in terms of the EEG data quality, additional artefacts and hardware solutions to remove these artefacts.

REFERENCES

[1] Sanei, S. & Chambers, J. A. *EEG Signal Processing.* Wiley Blackwell; 2007.

[2] Nunez, P. L. Physiological foundations of quantitative EEG analysis. In Tong, S. & Thankor, N. V. eds. *Quantitative EEG Analysis Methods and Clinical Applications.* Artech House; 2009.

[3] Gargiulo, G., Calvo, R. A., Bifulco, P. *et al.* A new EEG recording system for passive dry electrodes. *Clin Neurophysiol* 2010; **121**(5):686–93.

[4] Smith, S. J. EEG in the diagnosis, classification, and management of patients with epilepsy. *J Neurol Neurosurg Psychiatry* 2005; **76**(Suppl 2):2–7.

[5] Acharya, U. R., Sree, S. V., Swapna, G., Martis, R. J. & Suri, J. S. Automated EEG analysis of epilepsy: a review. *Knowledge-Based Systems* 2013; **45**:147–65.

[6] Katoh, T., Suzuki, A. & Ikeda, K. Electroencephalographic derivatives as a tool for predicting the depth of sedation and anesthesia induced by sevoflurane. *Anesthesiology* 1998; **88**(3):642–50.

[7] Morimoto, Y., Hagihira, S., Koizumi, Y. *et al.* The relationship between bispectral index and electroencephalographic parameters during isoflurane anesthesia. *Anesth Analg* 2004; **98**(5):1336–40.

[8] Sleigh, J. W., Andrzejowski, J., Steyn-Ross, A. & Steyn-Ross, M. The bispectral index: a measure of depth of sleep? *Anesth Analg* 1999; **88**(3):659–61.

[9] Jagadeesan, N., Wolfson, M., Chen, Y., Willingham, M. & Avidan, M. S. Brain monitoring during general anesthesia. *Trends Anaesth Crit Care* 2013; **3**:13–18.

[10] Brunner, P., Bianchi, L., Guger, C., Cincotti, F. & Schalk, G. Current trends in hardware and software for brain-computer interfaces (BCIs). *J Neural Eng* 2011; **8**(2):025001.

[11] Mak, J. N. & Wolpaw, J. R. Clinical applications of brain–computer interfaces: current state and future prospects. *IEEE Rev Biomed Eng* 2009; **2**:187–99.

[12] Moghimi, S., Kushki, A., Guerguerian, A. M. & Chau, T. A review of EEG-based brain-computer interfaces as access pathways for individuals with severe disabilities. *Assist Technol* 2013; **25**(2):99–110.

[13] Nicolas-Alonso, L. F. & Gomez-Gil, J. Brain–computer interfaces, a review. *Sensors* 2012; **12**(2):1211–79.

[14] Pasqualotto, E., Federici, S. & Belardinelli, M. O. Toward functioning and usable brain–computer interfaces (BCIs): a literature review. *Disabil Rehabil Assist Technol* 2012; **7**(2):89–103.

[15] King, C. E., Wang, P. T., McCrimmon, C. M. *et al.* The feasibility of a brain–computer interface functional electrical stimulation system for the restoration of overground walking after paraplegia. *J Neuroeng Rehabil* 2015; **12**:80.

[16] Do, A. H., Wang, P. T., King, C. E., Chun, S. N. & Nenadic, Z. Brain–computer interface controlled robotic gait orthosis. *J Neuroeng Rehabil* 2013; **10**:111.

[17] Ding, L., Worrell, G. A., Lagerlund, T. D. & He, B. Ictal source analysis: localization and imaging of causal interactions in humans. *Neuroimage* 2007; **34**(2):575–86.

[18] Kovac, S., Chaudhary, U. J., Rodionov, R. *et al.* Ictal EEG source imaging in frontal lobe epilepsy leads to improved lateralization compared with visual analysis. *J Clin Neurophysiol* 2014; **31**(1):10–20.

[19] Lu, Y., Yang, L., Worrell, G. A. *et al.* Dynamic imaging of seizure activity in pediatric epilepsy patients. *Clin Neurophysiol* 2012; **123**(11):2122–9.

[20] Sohrabpour, A., Lu, Y., Kankirawatana, P. *et al.* Effect of EEG electrode number on epileptic source localization in pediatric patients. *Clin Neurophysiol* 2015; **126**(3):472–80.

[21] Grech, R., Cassar, T., Muscat, J. *et al.* Review on solving the inverse problem in EEG source analysis. *J Neuroeng Rehabil* 2008; **5**:25.

[22] Birot, G., Spinelli, L., Vulliemoz, S. *et al.* Head model and electrical source imaging: a study of 38 epileptic patients. *Neuroimage Clin* 2014; **5**:77–83.

7 Digital Hearing Aids

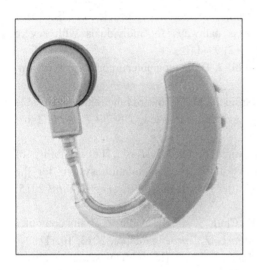

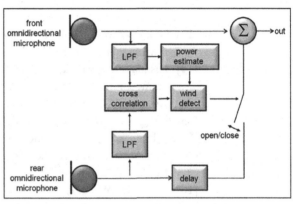

Introduction

Roughly 10% of the world's population suffers from some form of hearing loss, with over 5% or 360 million individuals suffering from hearing impairment that is considered to be disabling by World Health Organization (WHO) standards. The most common cause of hearing loss is damage to the cochlea, which results in reduced frequency selectivity and associated blurring of complex sounds. According to statistics compiled by the National Institute on Deafness and Other Communication Disorders (NIDCD) approximately 2 to 3 out of every 1000 children in the United States are born with a detectable level of hearing loss in one or both ears. One in eight people in the United States aged 12 years or older has hearing loss in both ears. Approximately 2% of adults aged 45 to 54, 8.5% aged 55 to 64, 25% aged 65 to 74, and 50% of those 75 and older have disabling hearing loss. Figure 7.1 shows the frequency and temporal characteristics of human speech, as well as the loss in different frequency ranges and classification of hearing loss from mild to profound.

Among people aged 70 and older who could benefit from hearing aids, only ~30% have ever used them, and the percentage for younger adults is significantly lower. The main reasons are the physical visibility of the device (related to social issues), the need for manual adjustments and problems in sub-optimal conditions such as strong winds and distracting background sounds (technical issues). In order to address the societal issue hearing aids are continually being designed to be smaller, with many of the miniaturization techniques outlined throughout this book enabling the reduction in size. Technical issues are being addressed via new developments in both hardware design as well as more powerful signal processing software. Each of these developments presents design challenges in terms of the requirement to package an ever-greater number of electronics in a more limited space, and to ensure that these electronics can operate in a power-efficient manner so as to maintain long battery life.

This chapter outlines the basic design of digital hearing aids, and the principles behind these designs [1]. The first sections in this chapter consider the physiology and acoustic properties of the human auditory system, the causes of hearing loss and the basics of hearing-aid design.

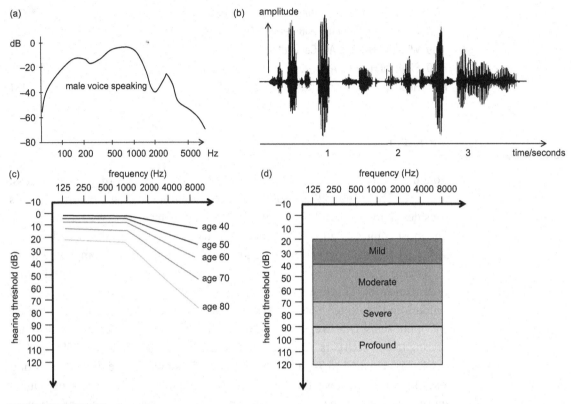

Figure 7.1 (a) Frequency spectrum of a male voice and (b) time plot of the same voice. (c) Plot of the hearing threshold of healthy individuals as a function of age. The decrease in high-frequency sensitivity is much greater than at lower frequencies. (d) Classification of mild to profound hearing loss in terms of the hearing threshold.

7.1 The Human Auditory System

Human hearing is most sensitive in the range of 1 kHz to 4 kHz. The structure of the auditory system is such that it enhances signals in this frequency range, and reduces the influences of both lower and higher frequencies. The basic structure of the human ear is divided into three parts: outer, middle and inner, as shown in Figure 7.2.

The pinna, also sometimes called the auricle, refers to the visible part of the ear that extends outside the head. As shown in Figure 7.2 the pinna collects sound and acts as a 'funnel' to amplify the sound and direct it towards the auditory canal. Left/right differentiation of sounds can be made since if sound comes from the right of a person it reaches the right pinna slightly before the left, and due to the 'shadowing' from the head the intensity is slightly higher as well. The pinna is highly structured, and also provides some degree of frequency discrimination,

Figure 7.2

Simplified (non-complete) schematic of the human auditory system. The pinna acts as a collector for sound, which then enters the ear canal. At the end of the ear canal the tympanic membrane vibrates due to acoustic waves. The cochlea converts sound to electric signals, and the hearing nerve conducts the signal to the brain.

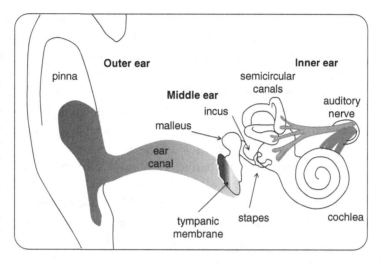

preferentially selecting those associated with human speech [2]. It performs this function by producing partial phase cancellation for certain frequencies in which the sound waves undergo different reflections from all of the small structures within the pinna. In the affected frequency band, called the pinna notch, the pinna effectively creates an acoustic band-stop or notch filter. For lower frequencies, corresponding to wavelengths much longer than the dimensions of the pinna structures, there is little attenuation of the signal. The notch frequency and attenuation also depend upon the vertical direction from which the sound comes, as shown in Figure 7.3(a).

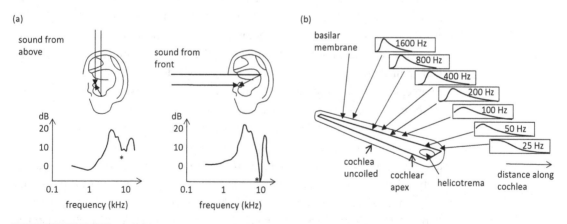

Figure 7.3   (a) Illustration of the different frequency-dependent amplification produced by the pinna as a function of the direction of the sound source with respect to the subject. The asterisk denotes the pinna notch frequency. (b) Plots of the basilar membrane frequency properties as a function of distance from the stapes.

From the pinna, the sound wave enters the ear canal, which essentially consists of an air-filled tubular structure that has a natural resonance at approximately 3.6 kHz and provides amplification between 1 and 5 kHz, the frequency range corresponding to human speech. The acoustic wave then passes through the tympanic membrane, which separates the outer and middle ear and converts the sound waves into a mechanical vibration. The mechanical motion is then transmitted into the cochlea, the first component of the inner ear. As shown in Figure 7.2, the cochlea is a spiral-like structure, which is filled with fluid. A structure called the basilar membrane is situated within the cochlea: this membrane tapers from a large diameter at the inlet of the cochlea to a smaller diameter at the end, as shown in Figure 7.3(b). The membrane can be mechanically displaced up and down by fluid flow. The resonant frequency of the membrane depends upon its diameter and stiffness: the larger the diameter the lower the frequency and the stiffer the membrane the higher the frequency. The membrane is widest and most flexible at its tip, and narrowest and stiffest at its base. Fluid movement within the cochlea therefore produces different resonance frequencies at different distances along the basilar membrane. The resonant frequency varies approximately logarithmically as a function of distance along the basilar membrane, as shown in Figure 7.3(b). This results in humans' logarithmic auditory perception, i.e. every doubling in the actual frequency is perceived as an octave increase in pitch.

There are many outer and inner hair cells located along the length of the basilar membrane. The upward motion of the basilar membrane causes electrical depolarization within the inner hair cells. These potentials feed into the auditory nerve and are transmitted to the brain. The outer hair cells alter their length in response to electrical feedback signals from the brain. This in turn alters the resonant properties of the basilar membrane. Together, the inner and outer hair cells act as a feedback loop: a larger displacement of the basilar membrane produced by a higher intensity sound wave produces a higher degree of depolarization and the perception of a loud sound in the brain. This signal feeds back to the outer hair cells to reduce the response of the basilar membrane.

Overall, the human auditory system transduces the sound pressure wave picked up by the pinna into an electric signal that is sent to the brain. The brain can interpret this information, together with visual cues such as who is talking, the direction from which the sound is coming (visually or by comparing the signal from both ears) and information about the frequency content of the particular sound, e.g. speech, music or machinery. In terms of reducing the effect of distracting noise, all of this information results in a very effective natural 'noise-suppression' circuit.

7.2 Causes of Hearing Loss

As first outlined in Chapter 2 there are two common types of hearing loss, **conductive loss** and **sensorineural loss**: patients may suffer from one or the other or both of these types. In conductive loss the middle ear is damaged. This condition is typically treated using a bone-anchored hearing aid that is implanted into the skull, and which transmits sound directly to the hearing nerve.

The more common sensorineural hearing loss affects patients who have damaged auditory nerves that are not able to send short electrical pulses to the brain.

The largest proportion of hearing impairment in the population relates to permanent sensorineural hearing loss associated with the natural aging process. The problem is getting greater due to increased longevity and the increased birth rates in many countries that occurred in the 1950s. In the coming decades, well over 1 billion people in the world will have age-related, permanent sensorineural hearing loss requiring hearing aids. Sensorineural loss can be treated either by hearing aids that magnify the sound signals, or cochlear implants that stimulate the nerve cells directly. Sensorineural hearing loss is characterized by:

(i) a reduced dynamic range between soft and loud sounds
(ii) the loss of certain frequency components (typically high frequencies), and
(iii) a reduced temporal resolution, i.e. the ability to detect signals of different frequencies that occur rapidly one after the other.

There are two important parameters, loudness and frequency, which characterize hearing loss. The human auditory system perceives sound on a logarithmic frequency and amplitude scale. The standard measure of loudness is the sound pressure level (SPL) defined as the amount in dB that a sound pressure wave exceeds the reference sound pressure, which is defined as 2×10^{-5} pascals, and is considered the threshold for human hearing.

$$SPL = 20\log \frac{sound\ pressure(Pa)}{2 \times 10^{-5}(Pa)} \tag{7.1}$$

Normal speech has an amplitude of approximately 60 dB but, as shown in Figure 7.1(d), the threshold for hearing in patients with even moderate hearing loss approaches this value. For 90% of adults requiring hearing aids the frequency response is lower than 60 dB between 500 Hz to 4 kHz.

The most common difficulty in understanding speech is the presence of background noise. This can be expressed quantitatively as the SNR-50, which is defined as the SNR for understanding 50% of speech. The difference in SNR-50 between normal hearing and patients with hearing loss is called the SNR-loss, which may be up to 30 dB, which means that people would have to speak 30 dB louder to maintain intelligibility! The major challenge in designing hearing aids is to reduce the effect of background noise and to increase speech intelligibility without introducing acoustic distortions.

7.3 Basic Design of a Digital Hearing Aid

The basic components of a digital hearing aid are an earmould, which fits into the ear and acts as a housing for the electrical circuitry; a microphone that converts sound pressure waves into an electrical signal; an amplifier, ADC, DSP board, DAC and receiver, which convert the filtered and amplified electrical signal back into a pressure wave. Modern microphones and receivers provide a flat response over a 16 kHz bandwidth, but significant amplification in most hearing aids can only be applied up to about 5 kHz [3]. The DSP performs three basic tasks: (i) multi-channel amplitude compression and (optional) frequency lowering to improve audibility; (ii) denoising, also referred to as 'sound cleaning', including reduction of background noise, acoustic feedback, wind noise, and intense transient sounds; and (iii) environment classification to enable the hearing aid to change its settings automatically for different listening situations: these topics are covered in detail in section 7.6. All components are powered by an in-built battery with a total drain current of approximately 1 mA, of which the microphone and ADC draw ~200 µA, the digital processor, memory unit and DAC ~700 µA, and the output amplifier and receiver ~50 µA. Digital hearing aids can be custom-programmed during the fitting process and have multiple listening profiles that can be selected by the patient. There are many features available on different models including volume control, wireless integration with smartphones, Bluetooth capabilities and wind-noise management. A schematic diagram of the components and their layout in a hearing aid is given in Figure 7.4

7.4 Different Styles of Hearing Aid

There are four main styles of hearing aids: behind the ear (BTE), in the ear (ITE), in the canal (ITC) and completely in the canal (CIC), each of which is shown in Figure 7.5. The BTE style sits behind the ear and has all the electronics housed outside the ear. The output from the receiver passes into the ear canal through a thin

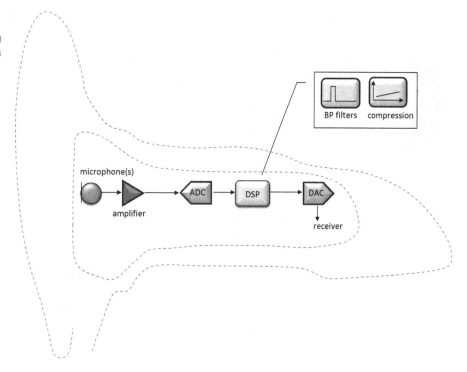

diameter tubing, which terminates in a patient-specific earmould. The ITE style
moves the hearing aid electronics into the outer ear by integrating them inside the
earmould. This style fills up most of the outer ear and appears as a solid mass.
The ITC design reduces the size further so that the earmould fits primarily into the
ear canal and reduces the space taken up in the outer ear, but it is still visible.
The CIC style is the smallest, fitting completely inside the ear canal, thus nearly
disappearing from view: a small string is used to remove the CIC from the ear.
The small size of the CIC and ITC means that there is only a small amount of room
for a receiver, and since a small receiver has a lower power output than a larger
one, the CIC and ITC hearing aids are most appropriate for patients with mild to
moderate hearing loss.

7.5 Components of a Hearing Aid

The individual elements of the general block diagram shown in Figure 7.4, as
well as earmoulds and associated vents, are described in detail in the following
sections.

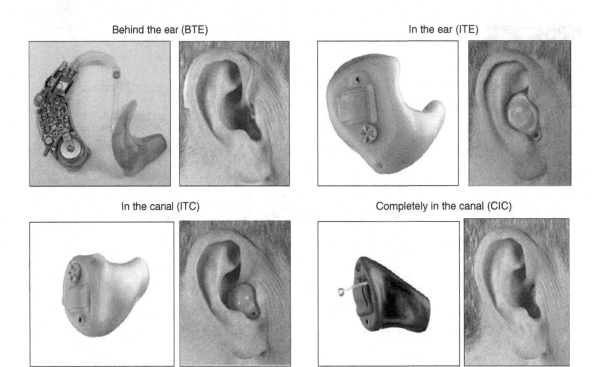

Behind the ear (BTE)

In the ear (ITE)

In the canal (ITC)

Completely in the canal (CIC)

Figure 7.5 Photographs of the four main styles of hearing aids and their appearances when in their respective positions.

7.5.1 Earmoulds and Vents

For the ITE, ITC and CIC systems, a clinician makes an ear impression from the patient and sends it to a manufacturer, who uses a casting technique to make a hollow plastic ear shell to fit inside the patient's ear canal. Several different materials can be used to make the impression including addition-cured or condensation-cured silicone, or acrylic. The shapes and positioning of the different types of earmoulds are shown in Figure 7.6. The microphone, amplifier, filters, ADC, DAC and receiver are all integrated inside the hollow earmould.

An important feature of the earmould is the vent, which is a hole drilled through the earmould. Without a vent the 'occlusion effect' occurs. This refers to the phenomenon of the bone-conducted sound vibrations of a person's own voice being trapped in the space between the tip of the earmould and the tympanic membrane, and then being reflected back towards the tympanic membrane via bone conduction and the flexing of the cartilaginous walls of the ear canal. The overall effect is to increase the loudness perception of the person's own voice. Compared to a completely open ear canal, the occlusion effect may increase the low frequency (below 500 Hz) sound pressure in the ear canal by 20 dB or

Figure 7.6

Schematics of three different types of hearing aid earmoulds (solid grey), and their positions with respect to the ear canal and tympanic membrane.

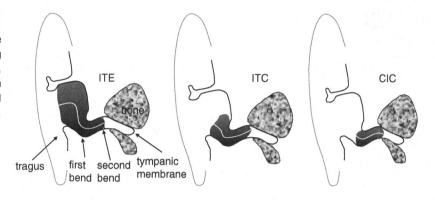

Figure 7.7

Sound can travel to the tympanic membrane via: (a) an amplified path through the microphone, amplifier and receiver; (b) a passive (unamplified) feed-forward path in which sound passes through the vent in the ear moulding; or (c) an active feedback path in which the sounds are re-amplified.

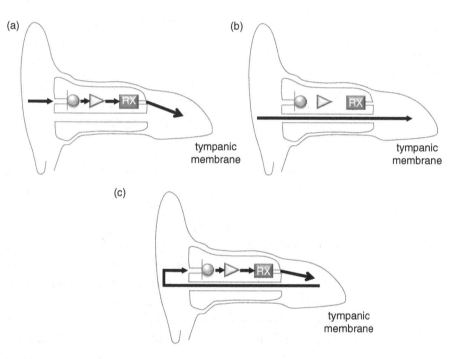

more. Having a vent in the earmould reduces the effect: the larger the diameter of the vent, the more the occlusion effect is reduced.

Vents do have some potentially detrimental effects as well, which are illustrated in Figure 7.7. The desired sound amplification path of the hearing aid is shown in Figure 7.7(a). However, a vent provides a secondary path of lower acoustic resistance through which unamplified sounds reach the tympanic membrane as shown in Figure 7.7(b). Indeed, as shown in Figure 7.7(c) a feedback loop can occur in which sounds that escape from the vent are picked up again by the

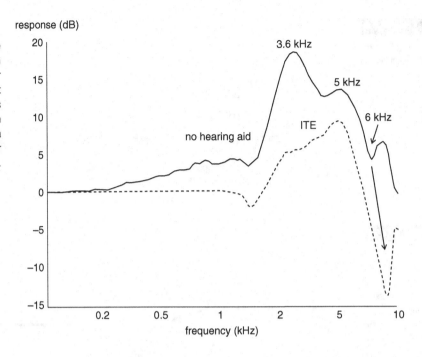

Figure 7.8

Schematic showing the effects of placing an ITE in the ear canal. The ear canal resonance at 3.6 kHz in the empty ear is not present with the ITE in place, and the 6 kHz pinna notch shifts to a higher frequency (see arrow).

microphone and re-amplified: this can cause acoustic oscillations that produce a feedback 'squeal'. This is particularly problematic for patients with severe hearing loss who require high-gain hearing aids. Various methods of reducing feedback are outlined in Section 7.6.1.

The effect of inserting an ITE earmould on the frequency spectrum of sound passing through the ear canal is shown in Figure 7.8. In the empty ear, peaks in the frequency spectrum occur at ~3.6 kHz (corresponding to the ear canal resonance, see later in this section) and 5 kHz, with a notch at ~6 kHz produced by the pinna as described previously. When an ITE fills the ear canal the effect is to remove the resonance feature at 3.6 kHz, and also to shift the 6 kHz notch to a higher frequency.

The term 'ear canal resonance' refers to the phenomenon in which the vent interacts with the residual volume of the ear canal to form a resonant system, representing an effective 'acoustic short circuit' at that frequency, which enhances the gain of the hearing aid. Analogous to the electrical circuits covered in Chapter 2, the resonance frequency (f_0) is given by:

$$f_0 = \frac{1}{2\pi} \frac{1}{\sqrt{M_{vent}(C_{canal} + C_{tymp})}} \tag{7.2}$$

where M_{vent} is the acoustic mass of the air in the vent, C_{canal} is the compliance of the air in the canal, and C_{tymp} is the compliance of the tympanic membrane. M_{vent} is given by:

$$M_{vent} = \frac{\rho_0 L}{\pi a^2} \tag{7.3}$$

where ρ_0 is the density of air, a is the radius of the vent, and L is its length. The compliance of the air in the canal can be expressed as:

$$C_{canal} = \frac{V}{\rho_0 c^2} \tag{7.4}$$

where V is the volume of air in the canal and c is the speed of sound (345 m/s). If one takes the very simple model of the open ear canal in a healthy subject being represented by an air-filled tube with length ~2.4 cm, then the resonance condition occurs at approximately 3.6 kHz (see Problem 7.7).

7.5.2 Microphones

The majority of microphone designs are based on electret technology, covered in section 2.6.2. Single microphones can be designed either to be omnidirectional, i.e. they detect signals from all directions with equal sensitivity, or directional, i.e. their sensitivity is higher for sounds from specific directions. Omnidirectional microphones have a frequency-independent sensitivity, whereas directional microphones have a lower sensitivity to lower frequencies. Dual-microphone set ups can be used to further enhance the directionality, and increasing the number of microphones in an array increases directional discrimination at the expense of increased hardware complexity and power consumption. These different types of microphone geometries are explained in the following sections.

7.5.2.1 Single Directional Microphones

Assuming that the signal that one wants to listen to comes predominantly from in front (as is the case for most conversations), the most simple method for reducing the influence of noise is to use a directional microphone that has a greater sensitivity for sounds coming from a particular direction, in this case directly in front of the listener. An omnidirectional microphone, such as the electret microphone outlined in Chapter 2, can be transformed into a first-order **directional microphone** simply by introducing a second sound tube, as shown in Figure 7.9.

In a directional microphone sound waves impinge on both sides of the diaphragm via plastic tubes at the front and rear of the microphone. There is a phase difference between waves either side of the diaphragm resulting from the

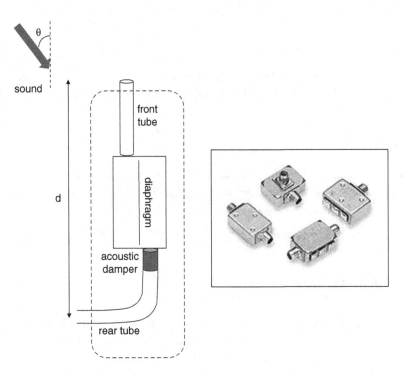

Figure 7.9

Schematic and photographs of a directional microphone. There are two sound tubes in this design, separated by a physical distance d.

difference in path length. The external phase difference, also termed the external phase delay, (ϕ_{ext}), is given by:

$$\phi_{ext}(^o) = 360\frac{d}{\lambda} \tag{7.5}$$

where d is the spacing between the front and rear tubes, and λ is the wavelength of the sound source. An acoustic damper is placed at the input to the rear port tube: this damper does not attenuate the sound wave, but is designed to produce an internal phase delay (ϕ_{int}) of the signal entering the rear port.

The detected signal (U) is a function of the incident angle θ of the sound source with respect to the microphone, as shown in Figure 7.9:

$$U(\theta) = (1 - k) + k\ cos\theta \tag{7.6}$$

where k, the delay ratio, is given by:

$$k = \frac{\phi_{ext}}{\phi_{ext} + \phi_{int}} \tag{7.7}$$

Consider a signal coming from directly behind the person wearing the hearing aid. The signal arrives at the rear port tube before the front port tube. If the phase delay φ_{int} is equal to φ_{ext} due to the physical separation of the two ports, then the signals from the front and rear tubes either side of the diaphragm have the same phase. This means that there is no net force on, and therefore no displacement of, the diaphragm. So a signal from the rear gives zero output from the microphone.

In contrast, any signal from the front will have a higher sensitivity, since the waves either side of the diaphragm are partially out of phase and so produce a displacement of the diaphragm. Therefore the microphone has a directionality, which can be quantified in terms of a directivity factor (DF):

$$DF = \frac{\text{power output of microphone for frontal signal source}}{\text{power output of microphone when signal is omnidirectional}} \quad (7.8)$$

This can be expressed in logarithmic form in terms of a directivity index (DI):

$$DI = 20\log DF \quad (7.9)$$

The standard way to display the directional dependence of the DI is on a polar plot, as shown in Figure 7.10(a). Concentric circles of constant attenuation ($-30, -20, -10$ and 0 dB) form one axis of the polar plot, with the other axis representing the angle of the source with respect to the microphone, which is located at the centre of the plot. Figure 7.10(b) shows polar plots corresponding to an omnidirectional microphone and Figure 7.10(c) to a directional one. Table 7.1 shows different polar patterns and their corresponding characteristics. These polar plots, of course, only correspond to a microphone in free space. The plots are altered by the fact that the microphone is placed inside the ear (see Problem 7.1).

One of the problems with a directional microphone is that lower frequencies are attenuated more than higher ones (see Problem 7.2). Figure 7.11 plots the sensitivity vs. frequency for an omnidirectional microphone and one operating in a mode that produces a cardioid pattern.

Although a few hearing aids still have directionality options based on a dual-port microphone, the incorporation of digital technology means that they have largely been replaced by dual directional microphones, described in the next section.

7.5.2.2 Dual Directional Microphones

The major limitation of the single directional microphone described in the previous section is that the polar pattern is fixed in the manufacturing process and cannot be altered afterwards. In order to avoid this limitation but maintain directionality, most modern hearing aids contain more than one microphone: these assemblies are termed microphone arrays and two examples are shown in Figure 7.12. The internal

Table 7.1 Delay ratio, null angle and directionality index corresponding to different polar sensitivity plots.

	Omni	Cardioid	Supercardioid	Hypercardioid	Figure-8
Polar pattern					
Delay ratio (k)	0	0.5	0.63	0.75	1.0
Null angle(s)	–	180°	±125°	±110°	±90°
DI	0 dB	4.8 dB	5.7 dB	6.0 dB	4.8 dB

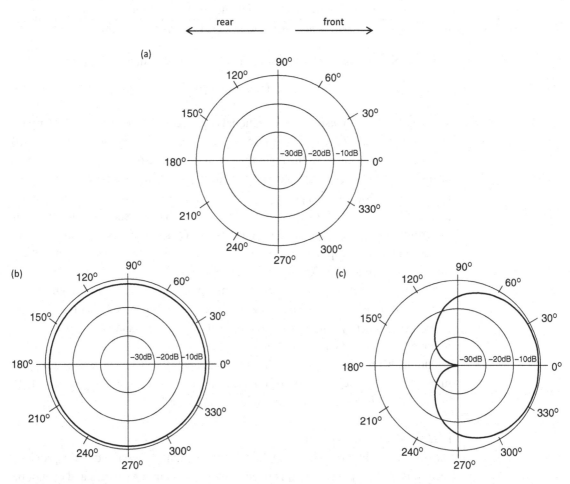

Figure 7.10 (a) Axes for a polar plot of the DI for a microphone. (b) Polar plot of an omnidirectional microphone. (c) Corresponding plot of a directional microphone, with the sensitivity highest towards the front of the microphone: this pattern is referred to as cardioid due to its heart-like appearance.

Figure 7.11

Comparison of the
frequency response of
a directional microphone
operating in
omnidirectional and
directional modes.

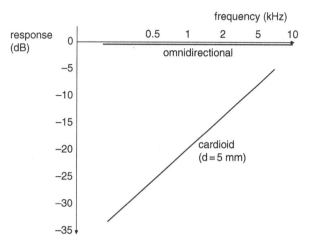

delay can be adapted on the fly by the DSP engine in the hearing aid and so different polar patterns can be produced for different listening situations. This process is termed 'beam-forming' [4]. As will be seen later in this chapter, after signal digitization the signal is split into different frequency bands for subsequent processing, and different delays can be implemented for different frequency bands, allowing frequency-adaptive beam-forming. The larger the number of microphones in the array, the higher the directionality, but this comes at the cost of greater size, higher cost and greater power consumption and so most systems currently consist of two separate omnidirectional microphones. Since low frequencies are attenuated when the microphone is operated in directional mode, a higher amplification factor must be applied at these lower frequencies. There are three main limitations of directional microphones.

(i) A higher sensitivity to wind noise. Wind noise is one of the largest problems for hearing-aid wearers. As the wind passes around the head it creates turbulence very close to the surface. Since directional microphones are very sensitive to sounds very close to the head, the wind noise in directional microphones can be substantially (20 to 30 dB) greater than in omnidirectional microphones. Since wind noise occurs predominantly at low frequencies, the effect is even greater when additional low-frequency gain is applied. Typically, the hearing aid is switched into omnidirectional mode in situations where wind noise is the dominant noise source. Some manufacturers also reduce the low frequency gain in this condition.

(ii) The higher internal noise relative to an omnidirectional microphone. The internal noise of a single omnidirectional microphone is roughly 28 dB SPL, and when two microphones are combined then the internal noise becomes

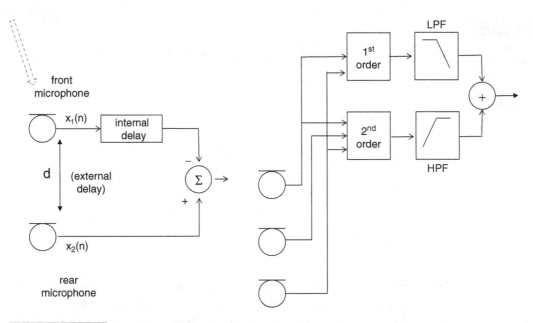

Figure 7.12 (Left) Schematic of a first-order dual directional microphone with each element having an independent output. (Right) Schematic of a second-order directional microphone with three elements. The outputs of the front and back elements form a first-order directional microphone, the output of which is fed into a low-pass filter. The outputs of all three elements are combined to form a second-order directional microphone, the output of which is high-pass filtered. These two filtered output signals are then summed together.

approximately 31 dB. Since low-frequency equalization is applied to overcome the low-frequency roll-off (shown in Figure 7.11), the internal microphone noise, which has a large component at low frequencies, is also amplified and may become problematic in very quiet environments. Partial solutions to this problem include using the omnidirectional mode in quiet situations, or not fully compensating for the low-frequency roll-off, i.e. increasing the gain by 3 dB/octave rather than the full 6 dB/octave.

(iii) The fact that very soft sounds from behind the person cannot be picked up.

One of the major manufacturing challenges for dual microphone systems is that the sensitivity and phase of the two omnidirectional microphones have to be extremely well matched, in the order of 0.01 dB in sensitivity and 1° in phase. For example, a directional microphone with an internal delay set to produce a hypercardioid pattern has an ideal DI of 6 dB with two nulls at approximately 110° and 250°. If there is a 1 dB mismatch in the sensitivities over the 1 to 1000 Hz range, then the directivity is reduced to 4.4 dB. Slight differences in phase have an even more dramatic effect: a 2° lag in the back microphone with respect to the front microphone changes the hypercardioid to a reverse hypercardioid pattern, where the null points to the front and the maximum sensitivity is now at the back! If there are very small residual errors after manufacture, a certain degree of microphone matching can also be performed using the DSP chip.

7.5.2.3 Second-Order Directional Microphone Arrays

First-order directional microphones typically improve the SNR of speech by 3 to 5 dB, but this number can be increased for patients with more severe hearing loss by increasing the number of microphones to three, as shown in Figure 7.12(b), in which case the system is referred to as second order. At low frequencies, the second-order frequency drop off is 12 dB/octave, so significantly higher than for the first-order mode. Therefore these types of microphone arrays most commonly operate in first-order mode at low frequencies and second-order mode at higher frequencies.

7.6   Digital Signal Processing

After the sound signal has been transduced by the particular microphone configuration into an electrical output, it is amplified and digitized by the ADC. The amplifier has a gain of between 10 and 30 dB and is usually integrated with the microphone. Although in principle hearing aid microphones can handle a dynamic range of 116 dB, in practice the microphone amplifier often clips or saturates for input sound levels above about 110 dB SPL because of limitations in the voltage that can be supplied by the hearing aid battery. The maximum peak–peak output voltage from a hearing aid microphone amplifier is typically 600 to 900 mV.

An anti-aliasing low-pass filter is used before the $\Delta\Sigma$ ADC as described in section 4.4.4.2. The $\Delta\Sigma$ ADC for a digital hearing aid typically has a sampling rate of 16 to 32 kHz, with 16-bit resolution, corresponding to 96 dB of dynamic range. However, in comparison with the 96 dB numerical dynamic range of the samples at the ADC's output, the ADC's input dynamic range may be reduced to only 80 to 85 dB to meet power consumption requirements [5]. The bandwidth is sufficient to cover the full range of speech and most music signals.

The digitized signals then pass to the DSP unit, which implements a number of different signal processing algorithms, discussed later in this chapter, which may include compression and clipping to increase listening comfort by reducing portions of the sound that are too loud, frequency shifting to translate speech to a lower frequency, which is helpful for people with high-frequency hearing loss, and wind-noise management that detects wind and eliminates the feedback that would otherwise cause ringing sounds to be heard [6]. A continuous time-

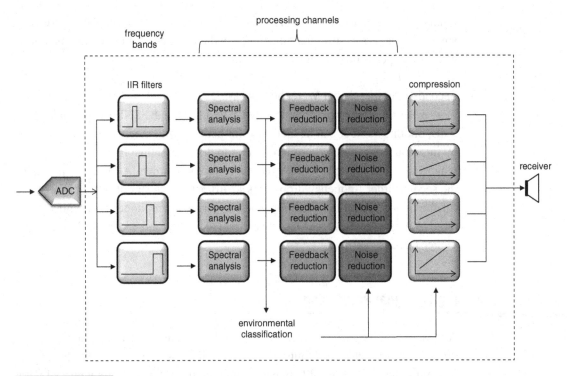

Figure 7.13 Schematic of a digital signal processing unit in which the signal is split into different frequency bands, each of which passes through a number of signal processing channels before being recombined at the output of the chip.

frequency analysis of the input signal is performed so that any changes in the sound characteristics can be matched by appropriate real-time changes in the processing parameters.

As shown in Figure 7.13 the DSP unit splits the signal into many different (typically 10 to 20) frequency bands, each of which can be processed differently. The channel bandwidth is smallest at low frequencies (80 to 150 Hz) and increases approximately proportional to the centre frequency [7]. The particular signal processing algorithms can either be hard wired into the DSP chip, which results in the lowest power consumption, or a more flexible programmable platform architecture can be used at the price of higher power consumption. Data is stored in erasable programmable read-only memory (EPROM), or can be temporarily stored in random access memory (RAM), which uses much less power. Some of the DSP code can be implemented in read-only memory (ROM), which has the lowest power consumption of all. One factor that must be considered is that the more sophisticated the algorithms used to process the signal the longer the time delay between sound arriving at the pinna and the signals being sent to the receiver and from there to the brain. If this delay can be kept to a few milliseconds then studies have shown that it does not interfere significantly with speech perception. However, if this delay is more than ~20 ms then a degree of discomfort is reported by patients: this effect is similar to a time delay between, for example, the video and audio on a skype call.

Digital signal processors in hearing aids typically run at clock speeds in the low tens of MHz range, much lower than in other signal processing engines such as computers and graphics cards, in order to reduce the power consumption. This limits the maximum number of computations that can be performed, and hence the complexity and number of algorithms that can run in real time. In general, DSP algorithms can be implemented in the time domain or frequency domain, or a hybrid of both. Time-domain processing effectively occurs immediately once the signal is received, whereas frequency-domain processing acts on a block of data, and therefore there is some delay between signal acquisition and processing. The advantages of frequency-domain processing are that there are fewer computations per sample required, which leads to lower power consumption.

Following the scheme shown in Figure 7.13, after passing through the IIR filters to produce the different frequency bands the signal is analyzed using either a time-domain filtering process or a transformation into the frequency domain. Both approaches involve a trade-off in time–frequency resolution. Spectral resolution should be sufficient for the subsequent signal-processing algorithms, but the time delay must be less than 10 ms, otherwise the sound echoes due to the interaction between the direct sound path through the vent and the delayed processed sound [8], as well as there being a mismatch between visual cues and the auditory signal as mentioned earlier.

Next, Figure 7.13 shows a module termed **environmental classification**, which essentially automatically selects which program (noise reduction, feedback cancellation, omnidirectional/directional mode, music/speech) should be used based on the patient's acoustic environment. Algorithms extract a variety of acoustic features and classify a listening environment by comparing the observed values of the features with a prestored map of values [9, 10]. Features such as the degree of synchrony between different frequency bands, amplitude differences between frequency bands and the estimated SNR in different frequency bands can be extracted in short time segments and evaluated over several seconds to identify changes in the acoustic environment. An important factor in making these types of algorithms successful is to mathematically 'train' the environment classifier using a wide variety of real-life sounds and situations.

7.6.1   Feedback Reduction

Acoustic feedback is most commonly encountered as a whistling sound produced by the hearing aid and tends to occur between 1.5 and 8 kHz. Feedback occurs from the component of the sound from the receiver, which escapes from the ear canal and is subsequently picked up by the microphone and re-amplified. The whistling sound occurs when the hearing aid oscillates, i.e. the amplification factor of the microphone is greater than the reduction in amplitude of the escaping sound, and therefore the feedback signal increases in magnitude. A smaller vent diameter can be used to reduce feedback but this also increases the occlusion effect, as discussed in section 7.5.1, and so new technologies concentrate on reducing feedback using DSP algorithms, so that a larger vent can be used to reduce the occlusion effect. There are three basic methods used to reduce the effects of feedback: (i) adaptive gain reduction, (ii) adaptive notch filters, and (iii) phase cancellation [11].

During the hearing aid fitting the clinician performs a series of tests to determine the maximum gain that can be used for low sound input levels for each frequency band to avoid feedback, and these values are programmed into the hearing aid. In **adaptive gain reduction**, the gain is reduced in the frequency band in which feedback occurs: the amount of gain reduction depends upon the level of feedback.

Adaptive notch filtering works by detecting the presence of feedback and generating a high-order notch filter at that frequency. This is done in narrow frequency bands to minimize reductions in speech intelligibility and sound quality. Fixed filters are set during the fitting procedure, but the adaptive component acts on abrupt changes in the acoustic environment [11].

Phase cancellation, as its name suggests, aims to generate a signal with a transfer function identical to the feedback signal, and then subtract this generated signal from the output of the microphone. The transfer function can potentially be derived by injecting a low-level noise signal into the receiver and then performing a cross-correlation of the input to the receiver with the output of the microphone: this procedure is performed when the hearing aid is fitted. Should feedback occur after this point, the characteristics of the digital filter can be changed in real time to account for the new feedback path, as illustrated in Figure 7.14.

7.6.2   Adaptive Directionality and Noise Reduction

Adaptive directional microphones use a dual microphone system, as covered earlier in this chapter. The internal delay can be dynamically changed by the DSP unit to

(a)

(b)

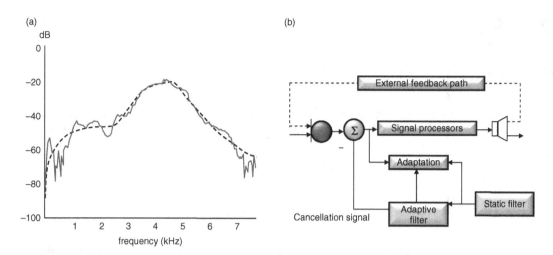

(a) Plot of the measured transfer function of a feedback signal (solid line) and that of the matched digital filter (dotted line). (b) Simplified schematic of a phase cancellation scheme to reduce feedback, which comprises both static and adaptive digital filtering.

produce different polar patterns, with the aim to adjust the maximum sensitivity to the desired sound source. In noisy environments, the direction of the sound source, e.g. in a group with more than one person talking, may change relatively rapidly and often. In directional mode, adaptive changes in polar pattern can be performed on the millisecond timescale. There are four basic steps that need to be performed by the DSP to implement adaptive directionality [12]:

(i) signal detection and analysis
(ii) determination of whether omnidirectional or directional mode should be implemented
(iii) calculation of the appropriate polar pattern, and
(iv) implemention of this pattern on the appropriate dynamic timescale.

Signal detection and analysis. There are several techniques used to detect the sound level entering the hearing aid. Most adaptive directional microphones only switch to directional mode when the sound level is above a prescribed value, remaining in omnidirectional mode in a quiet environment as described previously. In order to determine the relative amounts of real signal, i.e. speech, compared to the noise, a modulation detector can be used. The amplitude of speech signals has a constantly varying envelope with a modulation rate between ~2 and 40 Hz, with a peak at 5 Hz associated with the closing and opening of the vocal tract and the mouth. In contrast, noise usually has a relatively uniform level over a wide frequency range. The output of

Figure 7.15

Operation of a modulation detector to determine the SNR of a speech signal in the presence of noise. The solid line shows the signal maximum, and the thin line the signal minimum, which is assumed to be the noise level.

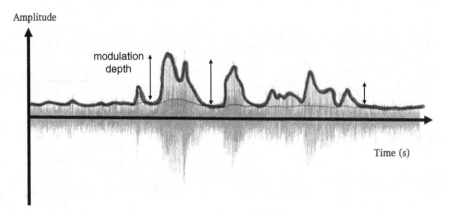

a modulation detector is shown in Figure 7.15, with the modulation depth calculated as a function of time.

Omnidirectional or directional mode, and appropriate polar pattern. As shown in Figure 7.15 the modulation detector measures the modulation depth of the signal centred at 4 to 6 Hz to estimate the SNR: the greater the modulation depth the higher the SNR. Depending upon the SNR value, the mode of the adaptive directional microphone can be switched between omnidirectional and directional mode. In directional mode, the appropriate polar pattern can be adaptively chosen by comparing the outputs of the two individual microphones. Some manufacturers implement this procedure on each of the different frequency bands in the DSP. Some adaptive directional microphone algorithms also incorporate a front/back detector in which the location of the dominant signals is determined. If a higher modulation depth is detected at the output of the back microphone the algorithm switches to (or remains in) the omnidirectional mode.

Appropriate dynamic timescale. The final parameter is the speed at which switching should take place, i.e. the adaptation/engaging times and release/disengaging times. Typical numbers for switching from omnidirectional to directional are 5 to 10 seconds, with the adaptation time to switch from one polar pattern to another being much faster. These time constants can be chosen at the fitting procedure to suit the particular desire of the patient.

7.6.3   Wind-Noise Reduction

The phenomenon of wind noise is familiar from everyday use of a mobile phone. Turbulent airflow over and around the microphone can cause significant noise, with SPL values up to 100 dB for a wind speed of ~5 metres per second in the case

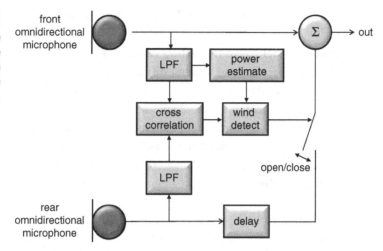

Figure 7.16

Block diagram of the circuit used to reduce the effect of wind noise in a dual-microphone hearing aid.

of a BTE hearing aid [13–15]. The noise level is maximum facing into the wind, and is higher at the low frequency end of the spectrum, with a maximum amplitude occurring at a frequency of several hundred hertz. As described previously, higher gain is applied at lower frequencies with a directional microphone array, which means that wind noise is amplified.

Essentially, when loud wind noise is detected, the hearing aid will either remain in, or be switched to, omnidirectional mode. Therefore the main issue is how to automatically detect whether indeed wind noise is present. Since the wind creates turbulence very close to the head, wind noise occurs in the near field, i.e. within 10 times the spacing between the two microphones, whereas speech occurs in the far field, which one can define as a distance more than 100 times greater than the microphone spacing. As a result, wind noise incident on the individual elements of a two-microphone array is uncorrelated, whereas the speech signal is highly correlated. The cross-correlation coefficient of the two signals can be calculated on the DSP chip: if the value is close to unity, then the two microphones are connected together to operate in directional mode; if the value is close to zero, then the rear one is disconnected, as shown in Figure 7.16, and the remaining microphone acts as a single omnidirectional device.

There are many other processing algorithms that can be used to determine the presence of wind noise (see Problems). Some hearing aid systems have different modes for different frequency bands, i.e. they can be omnidirectional at the lower frequency bands and directional in the upper frequency bands (this feature is referred to as split-directionality mode). If the decision is made to operate in the directional mode, then the appropriate polar pattern needs to be chosen. Most algorithms choose the pattern that adjusts the internal delay to minimize the output power (either for all frequencies, or for each frequency band individually).

7.6.4   Multi-Channel and Impulsive Noise-Reduction Algorithms

Multi-Channel Noise-Reduction Algorithms. These types of algorithm use the temporal separation and spectral differences between speech and noise to differentiate between the two. As described in the previous section they work by detecting modulation in the signal at 5 Hz. More sophisticated algorithms can also be used, for example using spectral intensity patterns across frequency bands, in which noise is assumed to be constant within and across bands, whereas speech is co-modulated by the fundamental frequency of the human voice. These algorithms are applied to each of the frequency bands that the signal is split into when it enters the DSP. The bands with a very low SNR are amplified by a small factor, and those with a high SNR by a correspondingly high gain. Some algorithms do not reduce the gain irrespective of the SNR unless the input level exceeds 50 to 60 dB, whereas others do not reduce the gain for frequencies between 1 and 2 kHz since this frequency range is so important for speech recognition. The more frequency bands that a device has, then the greater the potential noise reduction. However, there are longer processing times associated with a higher number of channels, and the total processing time cannot be more than ~10 ms.

Impulsive Noise Reduction. Sounds such as a door slamming have highly impulsive characteristics, which can cause significant annoyance to hearing-aid wearers. Even when fast-acting wide dynamic range compression is used, such impulsive sounds can be amplified excessively [16]. To avoid this, many hearing aids incorporate methods for detecting very fast increases in level and instantly apply a gain reduction for a short time in order to attenuate the signal.

7.6.5   Compression

In normal speech, there is an approximate 30 dB difference between the intensities of the loudest and softest sounds. The dynamic range of signals that can be processed by hearing-impaired patients is significantly reduced compared to healthy subjects, making speech much less intelligible. This means that the signals from the microphone with this 30 dB dynamic range must be **compressed** to have a lower dynamic range. Compression is also needed to reduce the loudness of very intense signals that could otherwise cause further damage to the auditory system.

As will be covered in the next section, the amount of required signal compression changes continuously as a function of time, corresponding to changes in the levels of normal conversation. Nevertheless, it is useful to consider the static

Figure 7.17

Characteristics of static compression as a function of sound input level.

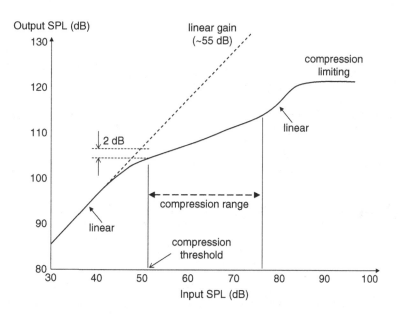

characteristics of a compression circuit at any one given time. Figure 7.17 shows a typical static compression plot of the output level as a function of input level, and an illustration of several terms used to characterize the compression graph. For very low input levels, up to about 40 dB SPL, the gain is either linear or more often actually increases with increasing level, i.e. the dynamic range is expanded rather than compressed. The **compression threshold** is defined as the input SPL above which compression begins: specifically, as the point at which the output deviates by 2 dB from linear amplification. The **compression ratio** is defined as the change in SPL input required to produce a 1 dB change in output level. The range of inputs over which compression occurs is termed the **compression range. Compression limiting**, seen at high input levels in Figure 7.17, means that there is a strong compression of the signal (>10:1) and effectively the output signal is independent of the input signal. The goal of output limiting is to ensure that very loud sounds are attenuated, while still creating as large a dynamic range as possible for sounds below this level. Compression can be performed either using a single circuit, or more than one circuit. For example, one could achieve the characteristics of Figure 7.17 using two different compressors: the first with a compression threshold of 52 dB SPL, a compression ratio of 3:1 and a compression range of 30 dB, and the second with a compression threshold of 87 dB SPL, a compression ratio of 10:1 and a compression range of greater than 15 dB.

Figure 7.18(a) shows a simple block diagram of a feedback compressor circuit. A fraction of the signal at the output of the amplifier is fed back through a level-detector circuit to an input that controls the amplifier gain. The control signal

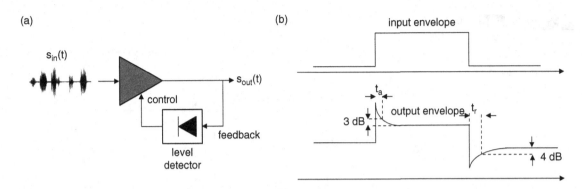

Figure 7.18 (a) Block diagram of the compression feedback loop used to implement either automatic volume level or automatic gain control. (b) Illustration of the effects of a sudden change in sound amplitude on the output from the hearing aid, with t_a representing the attack time, and t_r the recovery time.

therefore comprises a time-average of the output signal, and this signal controls the gain of the amplifier.

The feedback circuit has a time-constant that is controllable. If the time-constant of the feedback circuit is very long then the device is referred to as an automatic volume control (AVC); if the time-constant is very short then the circuit is referred to as an automatic gain control (AGC). The time-constant of the feedback loop when the signal increases in intensity, the so-called 'attack time' (t_a) shown in Figure 7.18(b), is defined as the time taken for the output to stabilize within 2 dB of its final level after the input increases from 55 to 80 dB SPL or alternatively within 3 dB of its final value after increasing from 55 to 90 dB SPL. The corresponding time-constant after a signal decreases in intensity is defined in terms of a 'recovery time' (t_r), also shown in Figure 7.18(b), corresponding to stabilization within 2 dB of its final value after the input decreases from 80 to 55 dB SPL or within 4 dB of its final value from 90 to 55 dB SPL.

In AVC systems the attack and recovery times are both between 0.5 and 20 seconds, much longer than typical sustained syllables in speech, which have durations between 150 and 200 ms. The advantages of **slow compression** using an AVC system include [17]:

(i) speech can be delivered at a comfortable level using a high compression ratio, regardless of the input level

(ii) the temporal envelope of speech is effectively undistorted, which is important for maintaining speech intelligibility

(iii) protection from intense brief transient sounds can be provided, and these transient sounds have only a very small effect on the long-term gain of the system, and

(iv) short-term level changes are preserved, so cues for sound localization based on left/right interaural level differences are not markedly disrupted.

Some of the disadvantages of having very long time-constants include:

(i) the systems may not deal very effectively with situations in which two voices with markedly different levels alternate with one another

(ii) when the user moves from a situation with high sound levels to one with lower levels (e.g. when leaving a noisy room), the gain takes some time to reach the value appropriate for the new situation. During this time the hearing aid may appear to become 'dead', and

(iii) when trying to listen to one (target) voice in the presence of another (background) voice, a normally hearing person can extract information about the target during the temporal dips in the background signal. This process is called listening in the dips. Long time-constants make this difficult because the gain does not increase significantly during brief dips in the input signal.

The alternative configuration, an AGC system, is much more fast acting, and is designed to make the patient's perception of the sound much closer to that of a normally hearing person. The attack times are typically between 0.5 and 20 ms, and the recovery times between 5 and 200 ms. These systems are often referred to as 'syllabic compressors' or as wide dynamic range compressors (WDRCs). These fast-acting syllabic compressors typically have lower compression ratios than AVC systems. The advantages of short time-constant circuitry include:

(i) it can, at least to a reasonable approximation, restore loudness perception to normal

(ii) it can restore the audibility of weak sounds rapidly following intense sounds, providing the potential for example for listening in the dips, and therefore performs well when two voices alternate with markedly different levels, and

(iii) it improves the ability to detect a weak consonant following a relatively intense vowel.

Some of the disadvantages include:

(i) when the input signal to the system is a mixture of voices from different talkers, fast-acting compression introduces 'cross-modulation' between the voices because the time-varying gain of the compressor is applied to the mixture of voices. This decreases the ability to distinguish between very different voices and leads to reduced speech intelligibility, and

(ii) under normal conditions in everyday life when moderate levels of background sound are present (e.g. noise from computer fans or ventilation systems), fast compression amplifies this noise.

From the considerations above it is clear that the ideal attack and response times depend upon the particular changes in intensity and frequency content of the input signal, and that fixed values represent a compromise. Therefore the concept of **adaptive release time** has been introduced, in which the release time is short (~20 ms) for brief but intense sounds, and becomes longer (~1 second) as the duration of the sound increases. One way of achieving this is a dual front-end compressor. This generates two compressed signals, one with long attack and recovery times and the other with shorter attack and recovery times. The slow control signal is the dominant one for normal operation; however, if there is a sudden increase in signal level (typically several dB above the time-averaged signal) then the faster acting control signal reduces the gain very quickly to avoid any discomfort to the patient.

7.6.6 Multi-Channel Compression: BILL and TILL

As shown in Figure 7.13, multi-channel hearing aids split the input signal into different frequency bands, which can then undergo different levels of compression since hearing loss may be related to specific frequency bands. If the microphone switches from omnidirectional to directional, there is also a change in the frequency characteristics, as shown in Figure 7.11, and this means that the compression levels in different frequency bands should also change. There are two general classes of multi-channel compression devices: BILL (bass increase at low levels) and TILL (treble increase at low levels). In BILL operation, low frequencies are preferentially amplified when the input intensity levels are low, and preferentially dampened when the input levels are high. For example, when input levels reach a given intensity at a given frequency (usually around 70 dB SPL at 500 Hz), the compression circuit automatically reduces the low-frequency amplification. The circuit is designed to assist people in functioning better in background noise. TILL is the exact opposite of BILL, i.e. preferentially amplifying the high frequencies at low input levels and dampening the high frequencies at high input levels. In this case, the monitoring frequency is about 2000 Hz. TILL is designed to prevent people from being bothered by high-frequency consonant sounds and to prevent speech from sounding tinny or sharp. An additional option is termed PILL (programmable increase at low levels), which can provide a combination of both BILL and TILL.

7.6.7 Frequency Lowering

People with profound hearing loss have particular difficulties hearing high frequencies, as illustrated in Figure 7.1(c). This means that the amplification required at these frequencies needs to be very high, which can lead to acoustic feedback or distortion, and potential damage to residual hearing. In addition, the dynamic range of such patients is typically very small, meaning that high gain often results in very loud sounds and accompanying discomfort. An alternative approach to using very high gain factors is **frequency lowering**, i.e. shifting high frequencies to lower ones. This process makes it easier to increase audibility while avoiding discomfort. There are many different ways of performing frequency lowering [18–20]. The most common one is termed **frequency compression**, which is an exact analogy of amplitude compression covered in the previous sections. Frequency compression does not occur over the entire frequency range, but rather frequency components up to a 'starting frequency', typically above 1.5 kHz, are not altered. Frequency components above this frequency are shifted downward by an amount that increases with increasing frequency. The advantage of this approach is the frequency lowered energy is not superimposed on energy from another frequency region.

7.7   Digital-to-Analogue Conversion and the Receiver

After all of the signal processing algorithms have been performed, the output signal from the DSP is converted back into an acoustic signal using a DAC and a receiver. The output data stream from the DSP is converted into a single-bit data stream at a much higher sampling frequency (~250 kHz). Loud sounds are coded as wide pulses while soft sounds produce narrow pulses in a technique known as pulse-width modulation, as shown in Figure 7.19. The single-bit data stream is then fed into the receiver via a class-D amplifier, also shown in Figure 7.19, which acts as a low-pass filter to produce the required analogue output. The advantage of this modulation technique is that the amplifier is only operational when there is an incoming signal present and so battery life is improved compared to a process in which the amplifier is always on.

The receiver converts the analogue electrical signal from the DAC into a sound wave, which travels to the tympanic membrane. The most common type of receiver uses balanced magnetic armatures as shown in Figure 7.20. The current from the DAC passes through a coil surrounding a flexible metallic armature (a structure that produces an electric current when it rotates in a magnetic field), thus magnetizing it. The armature is positioned between two permanent magnets, and thus begins to vibrate, which causes the drive pin to move. In turn this causes the diaphragm to vibrate, with the pressure created above the membrane creating acoustic waves that propagate through the sound

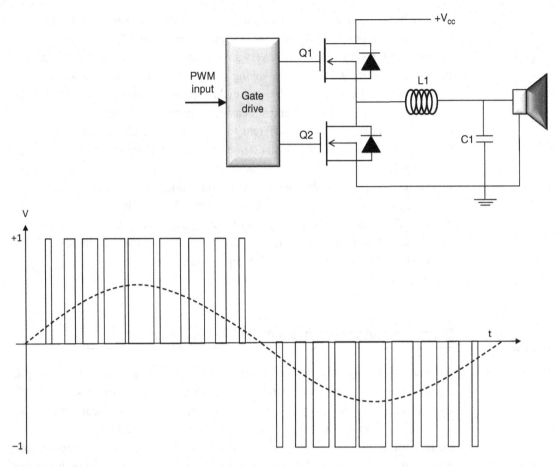

Figure 7.19   Schematic showing the principle of pulse-width modulation, as well as the basic circuitry for the amplifier. The inductor L1 and capacitor C1 form a low-pass filter at the output of the amplifier.

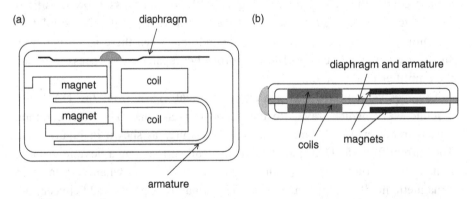

Figure 7.20   Two balanced-armature receivers. (a) Traditional construction, which uses a separate armature and diaphragm. (b) Flat motion construction, which combines the armature and diaphragm, allowing a total thickness of ~1 mm to be achieved.

outlet. The size of the receiver has decreased significantly over the years, with the thickness of a balanced armature device now being in the order of 1 mm. The frequency response of the receiver is non-uniform, with many peaks arising from wavelength effects within the sound tube, as well as mechanical resonances and the Helmholtz resonance mentioned previously. If not removed these effects would significantly distort the sound output and affect intelligibility, and so acoustic dampers are used to broaden out these peaks. There are many types of damper including small metallic particles, fine metallic meshes or materials such as plastic foam. The largest receivers can produce a peak output level of ~143 dB SPL, the smallest ~117 dB SPL: the larger receivers are still used for high-power hearing aids.

7.8 Power Requirements and Hearing Aid Batteries

Most hearing aids are powered by disposable zinc–air batteries, which have an output voltage of 1.4 volts and a current output over the battery's lifetime of approximately 300 mAh. Zinc–air batteries have high energy densities because the oxygen used in the reaction comes from the air and does not need to be part of the battery cell. Hearing aids are designed so that they will still operate when the voltage has dropped to 1.0 volts, but will sound an alarm at this level. The battery supplies power for the amplifier in the microphone, the DSP board and the receiver. There are four different standard battery sizes, each one suitable for a particular type of hearing aid: size 10 (5.8 mm wide by 3.6 mm high), size 312 (7.9 mm wide by 3.6 mm high), size 13 (7.9 mm wide by 5.4 mm high) and size 675 (11.6 mm wide by 5.4 mm high). When used for 16 hours per day, battery life ranges from a couple of days to a week, depending on the battery capacity and hearing-aid design. Rechargeable nickel metal hydride (NiMH) batteries do exist, but typically require charging every day. An accurate power gauge is critical to provide a warning so that the patient is not left with a non-functioning hearing aid.

7.9 Wireless Hearing Aid Connections

Wireless connectivity in hearing aids enables connection to external devices such as mobile phones, televisions and computers, as well as the microphones used in public address systems in theatres, concert halls, lecture halls, airports, train stations and bus stations. These types of direct connection increase the SNR of the acoustic signal, and significantly enhance the audibility and comfort for

Figure 7.21

(a) Photograph showing a telecoil inside a hearing aid. (b) Photograph of 'hearing glasses' in which a microphone array is placed along the arms of the glasses.

(a)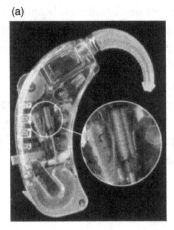

(b)

hearing-aid wearers. There are several different ways in which this wireless connectivity can be established.

For **electromagnetic transmission and reception**, the hearing aid contains a 'telecoil', which consists of an inductively coupled loop, coiled around a ferrite rod, as shown in Figure 7.21. Public places that transmit signals typically have a sign saying 'Hearing loop installed: switch hearing aid to T-coil'. The signal from the public-address system is fed into a large loop structure, which is contained in the structure of the building. Homes can also be fitted with this type of transmitter loop. The signal is transmitted at baseband, i.e. there is no carrier frequency, so the frequency range is approximately 1 to 20 kHz. Typically, there is a telecoil/microphone switch on the hearing aid, although in some models it is possible to have both operational simultaneously. Using a telecoil increases the SNR and, equally importantly, reduces the degree of room reverberation effects. A method that is used widely in classrooms in many educational facilities is to feed the audio signal (via a microphone) into an FM transmitter, which transmits the signal, usually at a frequency of 2.4 GHz. There is a corresponding FM receiver in a hearing-aid system, which can either be a universal component capable of attachment to many BTE models, or a proprietary integrated component compatible only with a particular model of BTE.

Another form of electromagnetic transmission and reception is termed **near-field magnetic induction** (NFMI), which works over very short distances of a couple of metres, and usually uses a loop worn around the neck as the transmitter, and a receiver very similar to a telecoil integrated into the hearing aid. The signal is modulated by a carrier frequency between 10 and 14 MHz. An audio signal from a phone/television/music player is plugged into the neck coil, and the signal wirelessly transmitted to the receiver in the hearing aid.

Direct streaming to the hearing aid can be achieved using smartphones that have Bluetooth Smart (a very low-power version of Bluetooth technology, also called Bluetooth 4.0, operating on bands around 2.4 GHz) capability. Direct streaming from other music devices and televisions is also possible using this technology, with significant improvements in SNR.

One of the implications of including wireless technology is that the battery has to power the wireless RF transmitter/receiver as well as the processor executing the transmission protocol. Because almost all 2.4 GHz wireless systems are digital and transmit their data in packets, there is an increased demand on the battery to be able to deliver fairly high current 'spikes' when transmitting these data packets. For this type of application, batteries with low internal resistance are required when wireless streaming is activated. As with all wireless devices the hearing aid must not create RF interference or result in excessive power being deposited in the body, and the hearing aid must be able to handle any interference received without causing any dangerous malfunctions. In the United States, for example, hearing aids have to be in compliance with FCC Directive CFR 47 Part 15, subpart C (more details on RF safety are given in Chapters 8 and 9).

When patients have more than 20 dB of SNR loss, the amplification available from regular hearing aids is not sufficient. One solution is to use head-worn or hand-held microphone arrays. Acoustic signals picked up by the microphone array are pre-processed and then sent to the hearing aid via the telecoil or an FM receiver. Using this approach 10 to 20 dB SNR gain can be achieved compared to a regular hearing aid. The array can be designed to operate in end-fire mode in which case the array can be integrated into the arm of a set of glasses, or a broadside array in which the array can be placed, for example, above a pair of glasses, as shown in Figure 7.21.

7.10 Binaural Hearing Aids

So far, this chapter has described the case of a hearing aid placed in only one ear. There are several potential advantages to having hearing aids in both ears: this arrangement is termed a **binaural hearing aid**. These advantages include minimizing the impact of signal attenuation due to the 'head shadow' effect, improved sound localization and increased dynamic range. From a practical point of view, however, having two hearing aids costs twice as much, and the two hearing aids must be wirelessly linked with a transmitter and receiver in each. This requirement for extra linking costs space and power, and can lead to extra time needed for signal processing. Patients who report the greatest benefit from binaural aids are those with more severe degrees of hearing impairment and those who spend considerable amounts of time in demanding and dynamic listening environments.

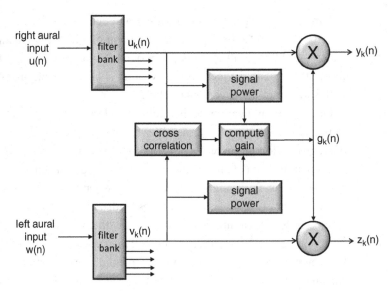

Using hearing aids in both ears rather than one ear increases the overall sound level by approximately a factor of 2 for soft sounds and a factor of 1.4 for loud signals. The pressure measured at two different locations is almost completely uncorrelated if the locations are greater than one-half the acoustic wavelength apart, e.g. 15 cm at a frequency of 1 kHz. A block diagram showing the basic signal processing for binaural hearing aids is shown in Figure 7.22. The four-microphone network forms left and right dual-microphone configurations. The first stage is to separate the right and left aural signals into different frequency bands, as described earlier in this chapter. The output signal from one side is sent to the contralateral side using wireless transmission. There it is processed together with the output signal of the ipsilateral dual microphone system, using a predefined weighting function. For each band, if there is a high correlation between the two signals from each ear then the gain is set to a high value. The overall gain function of the system is then modulated by the measured signal power to give the final outputs to the two aural systems.

One of the major advantages of binaural technology is its ability to make speech more intelligible in the presence of wind noise. Noise-reduction algorithms are based on the fact that if there is no correlation between the noise level and the phase of the output signals from the left and right ears, then wind can be assumed to be the cause. Additionally, the ratio between certain frequency components is analyzed to find out how strong the wind is. Within a few seconds, the two results trigger the particular noise suppression algorithm (essentially identical to those covered earlier in this chapter), depending on the level of wind noise. If the target signal level of the new wind-suppression system is higher than the level of the wind noise, the suppression will be reduced. This ensures that any speech components of the input signal are not adversely affected, thus increasing both comfort and intelligibility in windy situations.

An alternative approach to noise reduction is to use the 'better ear principle', in which one ear experiences a higher SNR due to the head-shadow effect. In this case, the signal with the higher SNR is transferred to the other ear and processed.

7.11   Hearing Aid Characterization Using KEMAR

In order to test new hearing-aid designs experimentally, and to characterize existing devices, an industry standard is required. The Knowles Electronics Manikin for Acoustic Research (KEMAR) Manikin Type 45 BA, shown in Figure 7.23, is the most commonly used model. The dimensions are based on the average male and female head and torso dimensions, and fulfil the requirements for the official testing standards ANSI S3.36/ASA58-1985 and IEC 60959:1990. The manikin is designed to simulate effects such as diffraction and reflection caused by the ears as sound waves pass the human head and torso. Different sized pinnae are also available, modelled over the range from small females to large males. A microphone is placed inside the manikin with an electrical output that can be hooked up to a recording device.

Figure 7.23

Photographs of the KEMAR phantom used for testing hearing aids. Figure reproduced with permission from Knowles.

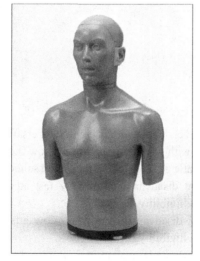
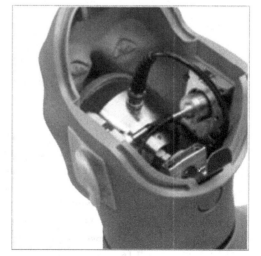

PROBLEMS

7.1 Sketch polar plots for an omnidirectional and a directional microphone placed inside the right ear of a patient.

7.2 In a simple directional microphone the sensitivity is lower for lower frequencies than for higher frequencies. Suggest possible reasons for this.

7.3 For a cardioid characteristic with a 10 mm spacing, what is the internal time delay in free space to cancel the external delay?

7.4 Draw the microphone direction patterns for the case when the time delay of the rear microphone with respect to the front microphone is: (a) $\tau = 2d/3\,c$, (b) $\tau = 0$.

7.5 Using first-order directional microphones, the directional effect at low frequencies reduces as the vent size increases. Suggest why this effect occurs.

7.6 Explain the frequency-dependent effects of increasing the tube diameter of the vent in an earmould.

7.7 Explain why a resonance condition occurs when the length of the air-filled ear canal equals one-quarter of the acoustic wavelength.

7.8 Starting from Figure 7.9 derive the expression for $U(\theta)$ in equation (7.6).

7.9 In Table 7.1 derive the null points for the supercardioid and hypercardioid polar patterns.

7.10 Suggest a situation in which frequency-dependent adaptive beam-forming would be useful.

7.11 Based on the static compression characteristic diagram shown in Figure 7.17, plot the static gain in dB as a function of input level.

7.12 Given the speech pattern shown below, plot the responses for the following attack and response times in terms of T_s: (i) $T_a = T_r = 10T_s$, (ii) $T_a = T_r = T_s$, (iii) $T_a = 0.1T_s$, $T_r = 0.3T_s$

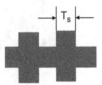

7.13 Draw typical signal levels for two different conversations, one with high dynamic range and one with low dynamic range. Show the effects of automatic volume level and automatic gain control on the output sound signals in each case.

7.14 Describe some potential disadvantages of very fast attack and release time constants for speech intelligibility.

7.15 Imagine some potential disadvantages of very slow attack and release time constants for speech intelligibility.

7.16 For the input signal $S_{in}(t)$ shown in Figure 7.18, plot the output signal $S_{out}(t)$ using (a) an AVC system, and (b) an ACG system.

7.17 There are many strategies for reducing the effect of wind noise by switching from directional to omnidirectional mode. One involves the differences between the outputs of the microphone in these two modes. (i) With the same low-frequency gain, which has the higher output for speech and wind sound? (ii) If the low-frequency gain is not equalized, how does the situation change?

7.18 Another method to detect wind noise can set different correlation criteria for the cross-correlation coefficients at low and high frequencies. Explain whether the coefficient is greater at high or low frequencies.

7.19 Describe possible effects of reflections off walls close to a patient who has a binaural hearing aid.

7.20 Explain why using a cell phone close to the ear can increase the level of feedback.

7.21 Feedback reduction methods can also potentially filter out musical sounds, since these are close to the single frequencies that characterize feedback. Come up with different concepts that might be used to differentiate between feedback and music.

7.22 The telecoil in a hearing aid is oriented parallel to the person's body, i.e. it is in the vertical direction. If you are designing a transmitting conducting loop that is placed in the walls and ceiling of a theatre, what orientation should this loop have and why?

7.23 Suppose that a loud sound slowly moves from in front of the subject to directly behind them. Describe what happens in terms of the directionality of the hearing aid, as well as the amplification and compression settings.

REFERENCES

[1] Kates, J. M. *Digital Hearing Aids*. Plural Publishing Inc; 2008.

[2] Batteau, D. W. Role of pinna in human localization. *Proc R Soc Ser B-Bio* 1967; **168**(1011):158.

[3] Aazh, H., Moore, B. C. & Prasher, D. The accuracy of matching target insertion gains with open-fit hearing aids. *Am J Audiol* 2012; **21**(2):175–80.

[4] Widrow, B. & Luo, F. L. Microphone arrays for hearing aids: an overview. *Speech Commun* 2003; **39**(1–2):139–46.

[5] Schmidt, M. Musicians and hearing aid design: is your hearing instrument being overworked? *Trends Amplif* 2012; **16**(3):140–5.

[6] Ricketts, T. A. Directional hearing aids: then and now. *J Rehabil Res Dev* 2005; **42**(4 Suppl 2):133–44.

[7] Glasberg, B. R. & Moore, B. C. Derivation of auditory filter shapes from notched-noise data. *Hear Res* 1990; **47**(1–2):103–38.

[8] Stone, M. A., Moore, B. C., Meisenbacher, K. & Derleth, R. P. Tolerable hearing aid delays. V. Estimation of limits for open canal fittings. *Ear Hear* 2008; **29**(4): 601–17.

[9] Nordqvist, P. & Leijon, A. An efficient robust sound classification algorithm for hearing aids. *J Acoust Soc Am* 2004; **115**(6):3033–41.

[10] Kates, J. M. Classification of background noises for hearing-aid applications. *J Acoust Soc Am* 1995; **97**(1):461–70.

[11] Chung, K. Challenges and recent developments in hearing aids. Part II. Feedback and occlusion effect reduction strategies, laser shell manufacturing processes, and other signal processing technologies. *Trends Amplif* 2004; **8**(4): 125–64.

[12] Chung, K. Challenges and recent developments in hearing aids. Part I. Speech understanding in noise, microphone technologies and noise reduction algorithms. *Trends Amplif* 2004; **8**(3):83–124.

[13] Chung, K. Wind noise in hearing aids: II. Effect of microphone directivity. *Int J Audiol* 2012; **51**(1):29–42.

[14] Chung, K. Wind noise in hearing aids: I. Effect of wide dynamic range compression and modulation-based noise reduction. *Int J Audiol* 2012; **51**(1):16–28.

[15] Chung, K. Effects of venting on wind noise levels measured at the eardrum. *Ear Hear* 2013; **34**(4):470–81.

[16] Moore, B. C. J., Fullgrabe, C. & Stone, M. A. Determination of preferred parameters for multichannel compression using individually fitted simulated hearing aids and paired comparisons. *Ear Hear* 2011; **32**(5):556–68.

[17] Moore, B. C. The choice of compression speed in hearing aids: theoretical and practical considerations and the role of individual differences. *Trends Amplif* 2008; **12**(2):103–12.

[18] Robinson, J. D., Baer, T. & Moore, B. C. Using transposition to improve consonant discrimination and detection for listeners with severe high-frequency hearing loss. *Int J Audiol* 2007; **46**(6):293–308.

[19] Simpson, A. Frequency-lowering devices for managing high-frequency hearing loss: a review. *Trends Amplif* 2009; **13**(2):87–106.

[20] Alexander, J. M. Individual variability in recognition of frequency-lowered speech. *Semin Hear* 2013; **34**:86–109.

8 Mobile Health, Wearable Health Technology and Wireless Implanted Devices

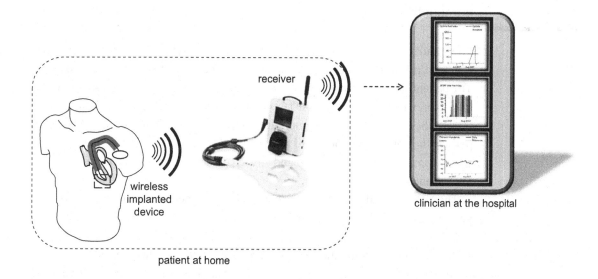

receiver

wireless implanted device

clinician at the hospital

patient at home

Introduction

The standard model of healthcare is that a patient with disease symptoms visits a general practitioner, who refers the case to a specialized physician at a hospital. The patient then visits the specialist who orders a battery of tests and scans. Some time later, when the results of the tests and scans have been interpreted, the patient revisits the hospital and the appropriate therapeutic course is prescribed. At regular periods the patient may be asked to revisit the hospital, whereupon updated tests are performed, progress assessed and any necessary changes to the medication are implemented.

There are several obvious inefficiencies in this process. First is the lack of any prior information that can be transmitted to the general practitioner before the patient's initial visit. Second is the requirement for the patient to visit the hospital multiple times. Perhaps the most important limitation is the lack of *continuous monitoring* of the patient's situation after therapy has started. Infrequent visits to the hospital provide only a snapshot of the entire treatment process. While the process may merely be inconvenient in the developed world, the entire concept breaks down in many developing countries.

This chapter considers the design of software and hardware that can be used to provide continuous information from a patient to a physician and vice versa, as well as from a device to the patient and the physician, all without the patient being required to be physically present at the hospital. Figure 8.1 schematically represents different ways in which information transfer can be achieved. In order of increasing complexity, the general areas covered are:

(i) the use of mobile/smartphone utilities for information dissemination and patient reminders: this comes under the general description of **mobile health**
(ii) the design of **wearable** activity monitors that can wirelessly upload data to the patient and physician

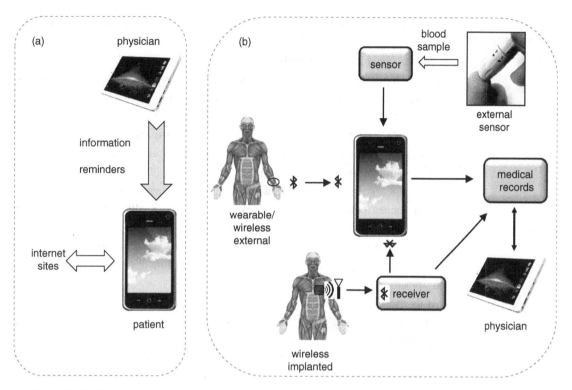

Figure 8.1 Illustration of the different modes of mobile and electronic health: (a) depicts the most simple example where the patient is contacted remotely and automatically by the physician to provide information or reminders for taking prescriptions; (b) illustrates different measurement sensors (external, wearable and implanted wireless), all of which send data to the physician. Information may be transferred via an app on a smartphone, or via a communications system that is intrinsically bound to the device in the case of an implanted wireless sensor.

(iii) the application of physical patches that can be attached to the skin and record heart rate, blood pressure, sweat, temperature etc. and can wirelessly send data to the hospital

(iv) the design and operation of **implanted monitoring devices** such as continuous glucose monitors, which can wirelessly transmit data through the body to a recording device that provides feedback to the patient and the physician, and

(v) the design of **implanted therapeutic devices** such as pacemakers and cardiac resynchronization therapy devices, which not only record and transmit data from patients who either have a history of heart failure or show signs of future heart conditions, but can also automatically provide therapeutic 'electrical shocks' when necessary.

8.1 Mobile and Electronic Health: Mobile Phones and Smartphone Apps

Mobile (m)-health was defined by the WHO in 2011 as 'medical and public health practice supported by mobile devices such as mobile phones, patient monitoring devices, personal digital assistants and other wireless devices'. Since this was written smartphone and tablet applications ('apps') have become increasingly dominant. In 2017 there are approximately 200,000 m-health apps, which operate on Apple and Android smartphones and associated tablets. The total commercial value is estimated to be approximately $26 billion in 2017, and is anticipated to rise substantially in the future. It should be noted that there is some ambiguity as to the differences between electronic (e)-health and m-health. Although one can simplify the situation to e-health referring to the reliance on electronic media such as the internet and computers, and m-health purely to those relying on mobile devices such as smartphones, there is no absolute line between the two descriptions.

At its most basic level, shown in Figure 8.1(a), m-health involves distributing healthcare information to patients over 3 G and 4 G mobile networks. For example, information has been very successfully sent automatically to pregnant women in Bangladesh, with feedback and questions sent in the other direction from the patient to the doctors. Doctors can set up automatic reminders for patients to take their medicines: even in 2017 up to 50% of medications are not taken according to the prescribed schedule, representing an enormous economic and health loss.

As another example, m-health has great potential in monitoring patients with mental health problems, particularly those who have psychotic episodes. Apps that use the embedded GPS in a smartphone can determine the patient's location to check that they have not strayed far from their homes or places that they normally frequent. Data from the in-built accelerometers provides information on physical activity. Microphones can detect the amount of interaction that the patient has with other people (the actual content of these conversations is not recorded). Sleep patterns can be inferred by the amount of time that the phone is not used and when the light sensors detect that it is dark.

One of the major motivations behind the growth in m-health is its very low cost: benefit ratio [1, 2]. For example, there have been a large number of studies that show positive feedback from patients when they monitor and record their own physical activity using devices with accelerometers. The amount of activity that patients undertake is increased significantly compared to patients who do not have this quantitative feedback. In one such study reported by Tang et al. [3] 415 patients with type 2 diabetes took part. The study included wirelessly uploaded home glucometer readings with graphical feedback; patient-specific diabetes summary status reports; nutrition and exercise logs; an insulin record; online messaging with the patient's health team, a nurse care manager and dietitian providing advice and medication management; and, finally, personalized text and video educational components sent electronically by the care team. The result was that the intervention group (compared to a control group) had significantly better control of their LDL cholesterol, lower treatment-distress scores and greater overall treatment satisfaction. As another example, in March 2012 the Dutch College of General Practitioners launched a non-commercial evidence-based public website in the Netherlands. Healthcare usage had decreased by 12% two years after providing this high-quality evidence-based online healthcare information.

Indeed, many of the challenges facing m-health are legal rather than medical. For example, can physicians be reimbursed for consultations over email rather than face to face? The FDA is responsible for the safety of medical devices, but if patients use devices to transfer their information to the hospital, is FDA approval necessary to check the accuracy of the data transfer process? Do apps that enable clinicians to see ECG data acquired remotely also need to be checked for accuracy and efficacy?

The rest of this chapter is concerned with one of the most rapidly developing elements of m-health, namely chronic disease management using remote monitoring devices. These enable patients to record their own health measures, which can then be sent directly to the hospital for examination. The devices covered include external wearable monitors, implanted wireless monitors, and implanted combined monitors and therapeutic devices.

8.2   Wearable Health Monitors

The boundary between medical devices and personal healthcare electronics is becoming increasingly blurred. For example, many major electronic and entertainment companies offer wearable devices that provide data such as the distance walked/run per day or an estimate of the number of hours slept. Within a clinical setting this data is very useful for a physician to be able to estimate how well the rehabilitation of a particular patient is progressing. Data from other sensors such as a blood pressure cuff, pulse oximeter or glucometer can also be acquired by the patients themselves. This data can be transmitted to physicians via wireless telecommunication devices, evaluated by the physician or via a clinical decision support algorithm, and the patient alerted if a problem is detected. As mentioned earlier, many studies have shown the improvement in patient outcome if remote patient monitoring (RPM) is used, as well as reported improvements in the quality of life and the interactions between patient and physician, reduction in the duration of hospital stays, mortality rate and costs to the healthcare system. Research has projected that in 2017 wearable wireless medical device sales will reach more than 100 million devices annually.

8.2.1 Technology for Wearable Sensors

The most common measurements performed by wearable sensors include heart rate, an ECG, surface temperature, blood pressure, blood oxygen saturation and physical activity. Much of the sensor technology used for wearable devices is very similar to that described in Chapter 2, with a particular emphasis on miniature sensors that are physically flexible. The major design differences compared to (larger and physically fixed) measurement equipment used in the hospital are that:

(i) the device must be able to be operated by a patient, i.e. a non-specialist, as opposed to a trained individual in a hospital. The main issue is usually poor data quality as a result of motion of the patient during data recording, and

(ii) data must be transferred wirelessly through a *secure network* to a recording device such as a smartphone, and from there to the physician.

A shortlist of the most common sensors used in wearable medical devices includes the following.

Accelerometers can be used in general to track the physical activity of patients, with particularly important applications to the time immediately post-operation in which sudden changes in mobility may occur. These are typically three-axis accelerometers, based upon piezoelectric or capacitive sensors. As an example, RPM sensors for patients with dementia who are susceptible to falls can either be worn by the patient or can be attached to their canes or walkers. GPS and accelerometer sensors monitor the patient's location, linear acceleration and angular velocity. They can detect sudden movements, and alert caregivers if the individual has fallen.

Pressure sensors, based also on piezoelectric, piezoresistive or capacitive sensors, are widely used for measuring heart and respiration rates. Capacitive sensors are also used to measure ECG signals and blood pressure. They can be integrated into clothing and household items, such as a bed or a chair.

Galvanic sensors can be used to measure the galvanic skin response (GSR), electrodermal response (EDR) or skin conductance (SC). These are measures

Figure 8.2

Different manifestations of wireless wearable devices: (a) and (b) are discrete sensors measuring parameters such as blood pressure, heart rate and galvanic skin response; (c) is an example of a wearable textile-based distributed sensor system; (d) shows a portable EEG headset with electrodes visible on the frame. Figures reproduced with permission from Emmosense, Emotiv and Isansys Corporations.

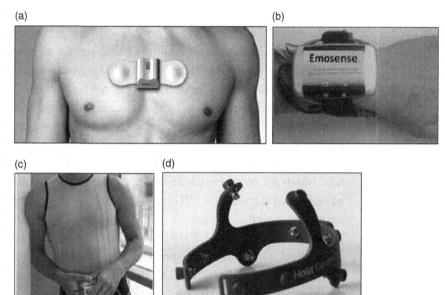

(a)   (b)

(c)   (d)

of the skin's electrical conductance (the inverse of electrical resistance), which varies with its moisture level. Galvanic sensors apply a very low voltage to the fingers and measure the current that flows through the skin. The perspiratory glands of the skin are controlled by the sympathetic nervous system, making their activity a good indicator for stress and activity.

Humidity sensors can be used to estimate parameters such as sweat rate via the water vapour pressure gradient measured by two humidity sensors at different distances from the skin.

Many of these individual sensors can be integrated into a single wearable device. Examples are shown in Figure 8.2, in which wearable battery-operated sensors can be designed as a stick-on patch or a device worn as a glove or wristband. The particular sensor shown in Figure 8.2(a) has a three-axis accelerometer, galvanic sensor and two temperature sensors (near body and skin). These sensors collect over 5000 readings from the body every minute, measuring the user's motion, skin temperature, heat flux and GSR, and can be used to determine the amount of time asleep, number of breaths per day and other related parameters. Data can be stored in the device's memory. The sensor wirelessly transmits a summary of the data to a smartphone, and from there to the physician's office. Sensors have also been embedded into objects with which the patient interacts, such as a chair, car seat and steering wheel, bed mattress, mirror, computer mouse, toilet seat and weighing scales [4].

Many of these sensors can be integrated into what are termed **smart textiles**. As an example, Figure 8.2(c) shows a T-shirt that measure the user's heart rate, respiration rate and skin temperature. Optical sensors similar to those used in the pulse oximeter have been integrated into clothing items such as gloves and hats to measure heart rate, blood pressure, respiration rate and SpO_2. Strain sensors such as piezoresistive films can easily be integrated into clothing and are used to measure respiration and heart rate. Inductive or piezoresistive sensors can be used to measure respiratory motion via changes in the coupling between two conductive coils.

Several companies have introduced wearable ECG necklaces and EEG headsets, which are not yet used medically but may be in the near future. The major advantage of the wireless EEG headset is that signals can be recorded while the subject is walking around. A commercial eight-electrode system shown in Figure 8.2(d) has a high CMRR (120 dB) and low noise floor (60 nV/√Hz). The system uses an ultra-low-power (200 μW) ASIC for the acquisition of the EEG signal and consumes less than 10 mW when sampling and streaming the data continuously at 1 kHz. The system is integrated into a box of size $5 \times 1 \times 1$ cm^3, which is packaged into an elastic headband.

As mentioned earlier, flexible electronic circuits also have enormous promise for wearable health applications [5]. Such circuits can even be applied directly to the skin to increase the SNR of the measurements. Examples are just beginning to be translated from the research laboratory into commercial applications. Figure 8.3 shows one example of a biocompatible tattoo [6–9], comprising three electrodes, to measure the concentration of lactic acid in sweat.

The final component of a wearable device is the transmitter needed to transfer data to a portable storage device. Data transfer from wearable devices to a smartphone or another receiver is usually via **Bluetooth**. Bluetooth technology operates within the frequency band of 2400 to 2483.5 MHz and uses frequency-hopping spread spectrum, i.e. the transmitted data is divided into packets and each packet is transmitted on one of the 79 designated Bluetooth channels, each of which has a bandwidth of 1 MHz. Data rates of several Mbits per second are possible using this technology. In addition, Bluetooth supports many security features, including password protection and encryption. There are many 'flavours' of Bluetooth including low-energy Bluetooth, which finds widespread use in implanted devices where low power consumption is a key element to ensuring long battery lifetime. In the future ZigBee may be a potential low-power alternative to Bluetooth, but has not been adopted so far to the same degree as Bluetooth. The data rates are slightly slower for ZigBee, although the number of devices that can be attached to a hub is much higher. ZigBee operates in the 906 to 924 MHz range (channels 1 to 10) and in the 2405 to 2480 MHz range (channels 11 to 26).

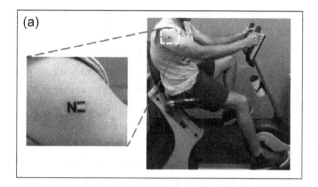

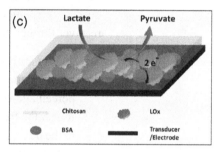

Figure 8.3   (a) An example of a tattoo, which acts as a sensor to measure lactate levels in sweat. (b) The electrochemical-based sensor consists of three electrodes, which are formed from chopped carbon fibres dispersed within both conductive carbon and silver/silver chloride inks to increase the tensile strength of the electrodes. (c) The chemical conversion of lactate to pyruvate releases electrons, which are detected as an electrical current. Figure reproduced from Jia *et al.* with permission from the American Chemical Society.

8.3   Design Considerations for Wireless Implanted Devices

The most common commercial implantable devices are cardiac defibrillators and pacemakers, followed by continuous glucose monitoring systems, which are covered in sections 8.4.1 and 8.4.2, respectively. There is also a large market for ingestible medical devices, which are swallowed by the patient, and then transmit endoscopic images through the abdomen [10]. These last types of devices are not covered in detail in this book, but essentially consist of a semiconductor-based image capture device, LEDs for illumination of the tissue and a video transceiver, all powered by a small internal battery.

8.3.1   Data Transmission Through the Body

If a device is surgically implanted into a patient, then one of the major design questions is how to transmit the data wirelessly from the sensor through the body to the receiver safely and with the highest SNR. Factors in the design process include choosing the frequency at which data is to be transmitted, the design of the antenna to transmit the data, the particular modulation method to be used for data transmission and whether the wireless implant should have an internal battery or can be powered by coupling to an external source. Each of these factors is considered in turn in the following sections.

8.3.1.1 Choice of Operating Frequency

After the data from the implanted sensor has been amplified and digitized it must be transmitted through the body to an external receiver. The choice of the frequency, or frequency range, at which to transmit the data from the sensor to the receiver through the body involves many factors:

(i) energy absorption by the (conductive) body increases with frequency. This has two important effects. First, the higher the frequency the lower the percentage of the transmitted signal that reaches the detector. Second, the higher the frequency the greater the amount of power deposited in tissue close to the transmitter, presenting a safety issue. The energy deposited is quantified in terms of a specific absorption rate (SAR), measured in watts/kg: there are strict federal limits on the allowable SAR for implanted medical devices (covered in the next section)

(ii) the size of the optimal antenna geometry decreases as a function of frequency, so it is easier to design small antennas at higher frequencies

(iii) the frequency must lie within one of the predefined government-approved ranges, discussed below and shown in Figure 8.4. Within each of the approved frequency ranges there are specified channel bandwidths, the value of which determines the maximum data transmission rate. The theoretical maximum channel capacity (C) measured in bits per second is related to the signal-to-noise (S/N) and channel bandwidth (BW) by:

$$C = BW \ln\left(1 + \frac{S}{N}\right) \tag{8.1}$$

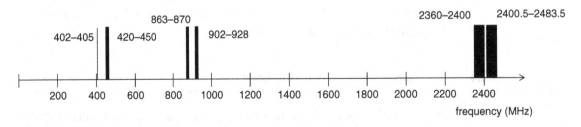

Figure 8.4   The major frequency bands (not to scale) used for the operation of wearable and implanted medical devices.

For example, for the operating frequency range of 402 to 405 MHz the bandwidth of each channel is less than 300 kHz, and so the maximum data transmission rate is relatively low, compared to the ultra-wideband (GHz) range where the channel bandwidth is much higher and therefore the data transfer rate can be much greater.

As shown in Figure 8.4, there are several government-approved frequency bands that can be used for medical devices, the most important of which are as follows.

(a) The **Medical Device Radiocommunication Service** (MDRS) mainly operates between 401 and 406 MHz, and has good RF penetration through the body at relatively low power levels. The main band, also referred to as the Medical Implant Communication Service (MICS) band, is 402 to 405 MHz; in addition, there are also two 'wing bands' between 401 to 402 MHz and 405 to 406 MHz (100 kHz bandwidth channels) for use by body-worn monitoring devices that do not communicate with implants.

(b) The **Wireless Medical Telemetry System** (WMTS) operates in three bands (608 to 614 MHz; 1395 to 1400 MHz and 1427 to 1429.5 MHz), and is used for 'measurement and recording of physiological parameters and other patient-related information via radiated bi- or uni-directional electromagnetic signals'. There are many disadvantages of the WMTS including interference in the lowest frequency band, and the limited bandwidth that can only support a few hundred signals. WMTS uses an unlicensed spectrum to communicate data from body sensors to remote monitoring locations.

(c) The **Industrial, Scientific and Medical** (ISM) frequency bands are used by many non-medical devices worldwide, but can also be used by wireless medical devices. Within these bands many methods have been used to minimize interference and data errors, but one issue is that data rates can be slowed down if there are other devices that use these bands. The most common frequency bands are 13.56 MHz, 418 to 433.9 MHz, 922 to 928 MHz and 2.402 to 2.483 GHz.

(d) The **Ultra-Wideband** range between 3.1 and 10.6 GHz, and with a bandwidth of at least 25% of the central frequency, can be used by devices that are able to detect motion, and therefore can be used to measure heart rate, respiration and patient body movement over a very short range (less than 1 m), but do not have to be in contact with the patient. To date only a very few UWB monitoring devices have been approved for use.

8.3.1.2 Antenna Design Considerations

The transmitting antenna integrated into an implanted device must ideally be small and flexible, have some degree of directionality towards the external receiver and

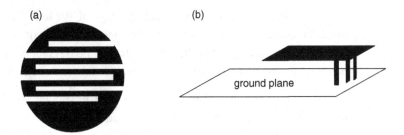

Figure 8.5 Examples of miniature antennas used for implantable medical devices. (a) An implantable patch antenna fabricated on a high permittivity substrate [11]. (b) An inverted-F antenna above a ground plane. Solid black areas represent metal conductors.

also have high gain. Many different conductor geometries can be used, including strips, loops and helices. The antenna is ideally resonated and impedance matched at the chosen operating frequency in order to maximize the signal transmitted through the body. Since the resonant frequency changes depending upon the environment, antennas are designed to be impedance matched over a wide bandwidth. In non-medical communications systems, such as radar or satellites, the length of an antenna is either one-half or one-quarter of a wavelength since this gives maximum sensitivity. However, at the frequency bands described in the previous section, these dimensions are much larger than the implanted medical device itself and therefore the size of the antenna has to be reduced while maintaining a high sensitivity. There are several different methods of miniaturizing the antenna including:

(i) using a high-permittivity substrate to reduce the effective electromagnetic wavelength (which is proportional to the square root of the permittivity) as shown in Figure 8.5(a)

(ii) using a spiral/helical structure to increase the effective length of an antenna while it still fits within a small volume

(iii) stacking small antennas together to form a composite antenna, or

(iv) using specialized structures such as inverted-F antennas, as shown in Figure 8.5(b).

The antenna must also be biocompatible in terms of the substrate on which it is formed, and this normally means that it is encapsulated in a polymer such as PDMS. Figure 8.5 shows different types of antenna that are used in implanted devices (very similar types of antenna are present in all smartphones).

In addition to maximizing the sensitivity of the antenna, the safety of such a device must also be considered. The power absorbed by the body, P_{abs}, from the antenna is proportional to the square of the electric field produced:

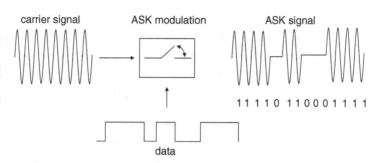

Figure 8.6

Principle of amplitude shift keying modulation in which a switch is turned on and off depending upon whether the input data consists of a zero or a one.

$$P_{abs} = \frac{1}{2} \int \sigma |E|^2 dV \qquad (8.2)$$

where σ is the tissue conductivity, the value of which increases with increasing frequency. Since antennas can produce strong local electric fields, the power output is limited to a value such that the SAR lies below the safety limits of 2 W/kg over 10 g of contiguous tissue (ICNIRP and IEEE standards).

8.3.1.3 Data Modulation Protocols

Having chosen the frequency and antenna design for data transmission, the next design choice is how the data should be modulated. Transmission should be as low power as possible, which in practice means that the modulation scheme should have a simple architecture, a high bandwidth to enable high data rates, and high accuracy in the presence of noise and distortion. It is also important to provide some form of data security.

Digital modulation techniques work by adding a digital signal onto a base carrier frequency. Different parameters such as the signal's amplitude, phase or frequency can be modulated. The most common modulation techniques used for biomedical devices are amplitude shift keying (ASK), phase shift keying (PSK) and frequency shift keying (FSK) [12]. The main criteria for choosing the modulation scheme are based on power, bandwidth and system efficiencies. The simplest modulation/demodulation scheme is ASK, shown in Figure 8.6, in which the amplitude of the transmitted signal is modulated by the digital output of the ADC of the implanted device to be either zero or one. In FSK modulation, two different frequencies are transmitted depending on whether the input is a zero or a one: this is referred to as binary FSK. Phase shift keying modulates the phase of the carrier signal according to the digital input: in the simplest scheme a zero input leaves the phase of the carrier signal the same, whereas a one inverts the signal by introducing a 180° phase shift. The relative advantages and disadvantages of the three different techniques can be summarized as:

ASK: simple modulation, low SNR, high error probability/low immunity to noise, high bandwidth efficiency and low power efficiency

FSK: high SNR, low error probability, requires twice the bandwidth of ASK, and

PSK: high SNR, low error probability, high power efficiency, low bandwidth efficiency and requires more complicated demodulation hardware.

There are many more sophisticated variations of the different schemes including quadrature phase shift keying, quadrature amplitude modulation and Gaussian minimum shift keying. Several specialized textbooks cover the advantages and disadvantages of these different techniques.

8.3.1.4 Wireless Powering of Implanted Devices

Power must be provided to the implanted device both for its intrinsic operation, i.e. for the sensor, amplifier and ADC, as well as for data transmission to an external receiver. Power can be provided in many different ways including via:

(i) a long-life conventional battery that is part of the implant

(ii) a rechargeable battery that can be charged contactlessly from outside the body

(iii) wireless transfer from an external source, or

(iv) energy harvesting from the body itself.

One example of a circuit that can externally power a medical implant is shown in Figure 8.7(a). The implant antenna is tuned to resonate at a certain frequency f_0 and absorbs power from the external antenna via mutual coupling. Within the implant, the AC signal at f_0 is rectified to a DC voltage to power components such as microcontrollers and pressure sensors, which are represented as a load capacitance and resistance. As covered in Chapter 2, a circuit that contains an inductor and a capacitor has a resonant frequency given by:

$$\omega_0 = \frac{1}{\sqrt{LC}} \tag{8.3}$$

If two circuits with identical or nearly identical resonant frequencies are placed close to one other, they couple together. The degree of coupling can be quantified via a coupling constant, k, which has values between 0 and 1, as shown in Figure 8.7(b).

The efficiency, U, of power transfer is given by:

$$U = k\sqrt{Q_1 Q_2} \tag{8.4}$$

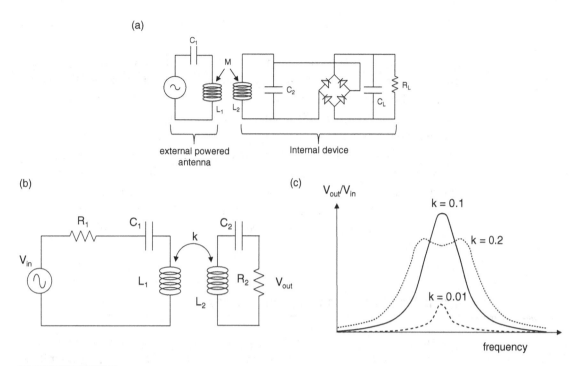

Figure 8.7   (a) An illustration of an external antenna coupling energy into a wireless implanted device, with diode-based rectification to produce a DC voltage. (b) Circuit diagram representing two coupled resonant circuits. (c) The voltage gain transfer function (and therefore the efficiency of power transfer) depends upon the value of the coupling coefficient, k, between the transmit and receive circuits. The maximum value corresponds to 'critical coupling', in this case k = 0.1.

where Q_1 and Q_2 are the quality factors of the external and internal circuits, respectively. The optimal power efficiency, η_{opt}, is given in terms of the value of U by:

$$\eta_{opt} = \frac{U^2}{\left(1 + \sqrt{1 + U^2}\right)^2} \tag{8.5}$$

The variation of the transfer function as a function of the coupling constant is shown in Figure 8.7(c). If the coupling is too small, e.g. the distance between the external and internal antennas is very large, then very little power is transferred and the system is referred to as undercoupled. If the two antennas are too close, then there is a frequency split in the resonances and, again, very little power is transferred: this situation is referred to as an overcoupled system. The ideal case is when the system is critically coupled, as illustrated in

Figure 8.7(c). For medical devices, typical coupling constants lie between 0.01 and 0.07.

Other ways of powering implanted devices include energy harvesting from the patient's movement: for example, using a piezoelectric-coated film or fibres woven into a compression jersey, every time that movement occurs and the film expands a small current is generated. Another source of internal power is body heat: one possibility is to utilize the Seebeck effect, in which a thermal gradient between two sensors can be used to generate a voltage. The energy produced by internal harvesting can be stored in an internal battery to power the sensor and antenna when needed for data transmission. Whichever method is used it is important to be able to generate sufficient power: tens to hundreds of microwatts are required to run the sensor alone.

8.4 Examples of Wireless Implanted Devices

As stated previously, the largest category of wireless implanted devices is cardiovascular based, followed by continuous glucose monitors. The market for implanted devices is expanding rapidly and most of the major medical device companies (including Abbott, Johnson & Johnson, Medtronic, Baxter, Stryker, Boston Scientific and Bayer) have significant R&D and commercial interests in this field. In terms of cardiovascular devices, the most common are pacemakers, defibrillators and resynchronization therapy devices, each of which is covered in detail in the following sections.

8.4.1 Cardiovascular Implantable Electronic Devices

The number of patients with cardiovascular implantable electronic devices (CIEDs) has increased very rapidly over the years. In the developed world, acute decompensated heart failure (ADHF), the congestion of multiple organs by fluid that is not drained properly by the heart, is the leading cause of hospitalization among patients over 65 years of age. Hospital post-operation re-admission rates are on the order of 50 to 60%. Many of the symptoms of ADHF can potentially be detected by continuous monitoring of various heart functions and measures, and studies have shown a significant and clinically

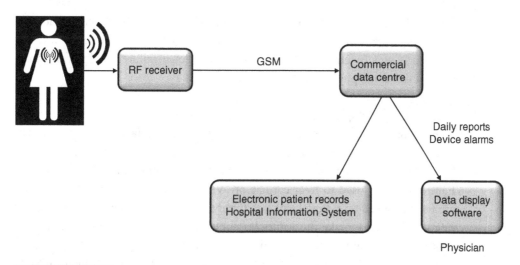

Figure 8.8   Data flow from a cardiovascular implanted electronic device to the physician.

important reduction in ADHF episodes using an implantable sensor-based strategy [13]. The major types of CIED are:

(i)   pacemakers and implantable cardioverter defibrillators (ICDs)
(ii) cardiac resynchronization therapy (CRT) devices
(iii) implantable loop recorders (ILRs), and
(iv) implantable haemodynamic monitors (IHMs).

A CIED has a small antenna that transmits data through the body to an external receiver. The antenna normally operates in a frequency range between 402 and 405 MHz and is typically a variant of the planar inverted-F antenna (PIFA) discussed in section 8.3.1.2. Although the device collects data continuously, the antenna is usually programmed to send data only once a day via GSM technology to the hospital's central database, as shown in Figure 8.8. Since 2015, several companies have provided apps on mobile phones (both iphone and Android based) that can automatically transmit data to the physician.

The wireless connection between the implanted device and the external transmitter/receiver can also be used to perform remote monitoring to detect any device failure. The major device failure in CIEDs is lead failure. There can be significant mechanical stresses on these leads, and current failure rates over five years have been reported to be in the 2 to 15% range, depending upon the particular device.

The CIED performs a periodic measurement of the impedance of each lead, and this value is transmitted wirelessly to the external receiver to determine whether the lead is still functioning. In the case of suspected failure this information is transmitted to the physician.

8.4.1.1 Pacemakers and Implanted Cardioverter Defibrillators

Approximately 200 000 patients receive pacemaker implants every year in the United States. Dual chamber pacemakers represent over 80%, with the remainder being single ventricular (SVI), single-atrial or biventricular (BiV) devices. Pacemakers are used to treat cardiac arrhythmias as well as heart block. They are used to speed up a slow heart rate (bradycardia), slow down a too rapid heart rate (tachycardia), make the ventricles contract normally when the atria are not contracting normally (atrial fibrillation) or ensure the passage of electrical current from the upper to the lower chambers of the heart in the case where there is a conduction block between the two chambers (heart block). Pacemakers also monitor and record the heart's electrical activity and heart rhythm. Adaptable pacemakers can monitor blood temperature, breathing rate and other factors, and adjust the paced heart rate according to these parameters. For example, a patient who is exercising will naturally have a higher heart rate than one who is at rest, and the algorithm to detect and correct for tachycardia needs to be able to take this into account.

A pacemaker applies electrical pulses to force the heart to beat normally. It is a device a few centimetres in size that is implanted into the chest, as shown in Figure 8.9(a). Pacemaker implantation is performed by first guiding the pacemaker lead(s) through a vein and into a chamber of the heart, where the lead is secured. The pacemaker is then inserted into a surgically formed pocket just above the upper abdominal quadrant. The lead wire is connected to the pacemaker, and the pocket is sewn shut. This operation can now be performed on an outpatient basis, and eliminates any need for open-chest surgery. The lead is made of a metal alloy with a plastic sheath so that the metal tip of the lead is exposed, as shown in Figure 8.9(b). Many of the leads have a screw-in tip, which helps anchor them to the inner wall of the heart. A pacemaker battery (normally lithium/iodide or cadmium/nickel oxide) provides power to stimulate the heart, as well as providing power to the sensors and timing devices.

The pacemaker electrodes detect the heart's electrical activity and send data through the wires to the computer in the pacemaker generator. Electrical pulses travel through the same wires to the heart. Pacemakers have between one and three wires that are each placed in different chambers of the heart. The wires in a single-chamber pacemaker usually transmit electrical pulses to the right ventricle. The two wires in a dual-chamber (DDD) pacemaker transmit electrical pulses to

(a)

(b)

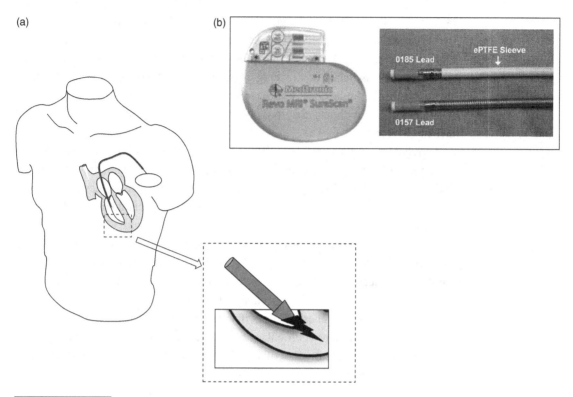

Figure 8.9 (a) Schematic of a single lead pacemaker surgically placed in a patient. The wire with the electrode is inserted into the heart's right ventricle through a vein in the upper chest. The insert shows the electrode providing electrical stimulation to the heart muscle. (b) Photographs of a commercial pacemaker, as well as two different types of pacemaker leads. Permission obtained from Medtronic Inc.

the right atrium and the right ventricle in order to coordinate the contractile motions of the two chambers. The two chambers can be paced or inhibited depending on the measured cardiac activity. One of the major advantages of the DDD pacemaker is that it preserves AV synchrony. If the pacemaker does not sense any intrinsic atrial activity after a preset interval, it generates an atrial stimulus and initiates a time period known as the AV interval. During this time period the atrial channel of the pacemaker is non-active. At the end of the time period, if no intrinsic ventricular activity has been sensed by the ventricular channel, the pacemaker generates a ventricular stimulus. Following this stimulus, the atrial channel remains non-active for a short period termed the post-ventricular atrial refractory period (PVARP) to prevent possible interpretation of the ventricular stimulus or resulting retrograde P waves as intrinsic atrial activity. The total atrial refractory period (TARP) is defined as the sum of the AV and PVARP intervals. Most DDD pacemakers operate in switching mode. If a patient develops supraventricular tachycardia, the pacemaker might pace the ventricles at a rapid rate based on the atrial stimulus. To prevent this situation from developing a mode-switching algorithm is used, whereby if a patient develops an atrial tachydysrhythmia, the pacemaker switches to a pacing mode in which there is no atrial tracking. When the dysrhythmia stops, the pacemaker reverts to DDD mode, thus restoring AV synchrony without the possibility of inadvertently supporting atrial tachydysrhythmias.

A biventricular pacemaker has three wires, which transmit pulses between one atrium and both ventricles in order to coordinate the electrical activity of the left and right ventricles. This type of pacemaker is also called a CRT device (see next section). Block diagrams of an adaptive rate-response pacemaker and the circuit used to produce pulses of electric current are shown in Figure 8.10.

In the case of more serious cardiac arrhythmias, which are considered life threatening, an ICD may be surgically implanted into the patient. An ICD is similar to a pacemaker in that it can treat abnormally slow heart rhythms with electrical pulses of relatively low energy. However, it also has the capability of providing much higher energy pulses if severely abnormal heart rhythms such as ventricular tachycardia or ventricular fibrillation are detected. Defibrillation works by depolarizing a critical mass of the heart muscle, enabling the normal sinus rhythm to be re-established by the pacemaker cells in the SA node.

8.4.1.2 Cardiac Resynchronization Therapy Devices

Cardiac resynchronization therapy (CRT) is used to treat heart failure, and can either be administered using a special pacemaker (a CRT-P), or an ICD with bradycardia pacing capabilities, called a CRT-D. The CRT is implanted in exactly the same way as a standard pacemaker or ICD, except that a third pacing lead is added to stimulate the heart. The device can be programmed externally and stores data, which can be transmitted wirelessly to a receiver.

Some ICDs or CRTs can also measure intrathoracic impedance, which is the impedance between the right ventricular lead and the device housing. The impedance value reflects the amount of fluid accumulation around the heart and lungs [15] and this impedance has been shown to decrease approximately 15 days before the onset of heart failure, thus providing an 'early warning system' for the clinician. The measurement of intrathoracic impedance is based on the principle that water is a relatively good conductor of electrical current. A very low electric current is produced by the right ventricular lead at a frequency of 16 Hz and passes through the thorax to the metal can of the ICD or CRT, as shown in Figure 8.11. The potential difference between the current source and the ICD/CRT is measured and the impedance calculated from this value. The water content is known to be inversely related to the measured impedance, and so a steadily decreasing

(a)

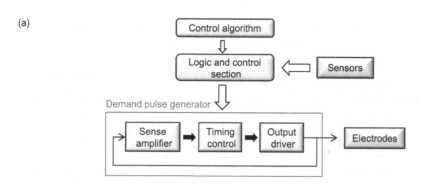

(b)

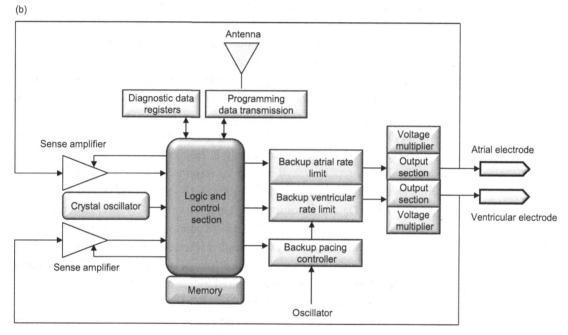

Figure 8.10 (a) Block diagram of a rate-responsive pacemaker. (b) Block diagram of an electric pulse generator. Figures adapted from reference [14].

Figure 8.11

(a) Schematic showing the measurement of intrathoracic impedance, and (b) the impedance changes when fluid accumulates in the lungs and around the heart.

(a)　　　(b)

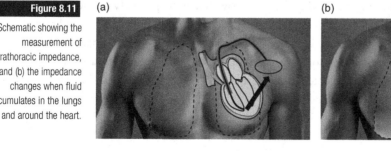

impedance is a sign of increased fluid accumulation. These measurements are taken several times a day to determine whether the impedance shows a decreasing trend, which may require patient hospitalization.

In addition to active devices that can be used to try to control heart function, there are a number of implanted devices that are used solely to monitor heart function. These are covered in the next sections.

8.4.1.3   Implanted Loop Recorders and Haemodynamic Monitors

An ILR is used to record long-term (over many years) heart signals. It is a subcutaneously implanted ECG monitoring device, with a rectangular geometry of roughly $60 \times 20 \times 5$ mm, with two built-in leads on the case of the device, which are used to acquire a single-lead bipolar ECG trace. It is used for diagnosis in patients with recurrent unexplained episodes of heart palpitations or syncope (a transient loss of consciousness with spontaneous recovery), for long-term monitoring in patients at risk for, or with already documented, atrial fibrillation (AF) and for risk stratification in patients who have sustained a myocardial infarction (MI). The device is typically implanted in the left pectoral region and stores single-lead ECG data automatically in response to detection of significant bradyarrhythmia or tachyarrhythmia, or in response to patient activation via an external hand-held antenna should the patient feel light-headed or lose consciousness temporarily. Data is transmitted wirelessly immediately after collection, and then new data is recorded over the old data in a continual 'loop', hence the name of the device. An ILR is particularly useful either when symptoms are infrequent (and thus not amenable to diagnosis using short-term external ECG recording techniques) or when long-term trends are required by the physician.

An implantable haemodynamic monitor (IHM) is also a single-lead device, which can measure systolic and diastolic pressure, estimate pulmonary artery diastolic pressure, right ventricular (RV) changes in pressure, heart rate and core body temperature. The lead is implanted in the right ventricular outflow tract with a tip electrode at the distal end. ECG and heart rate data is acquired via a unipolar vector between the tip electrode and the IHM casing. There is also a pressure sensor, used to measure right ventricular pressure, located a few centimetres away from the tip electrode, placed below the tricuspid valve. The pressure sensor capsule is hermetically sealed and consists of a deflectable diaphragm, which measures capacitance changes caused by changes in RV pressure. Data can be continuously monitored, or data collection can be triggered by events such as automatically detected bradycardia or tachycardia, or can be patient initiated similarly to the ILR.

8.4.1.4 Implanted Pulmonary Artery Pressure Sensor

The pulmonary artery (PA) carries deoxygenated blood from the right ventricle to the lungs. The pressure in this artery in normally between 9 and 12 mmHg, but in conditions such as pulmonary artery hypertension the pressure increases dramatically to above 25 mmHg. This condition is related to heart failure or severe lung/airway diseases. In order to measure the PA pressure continuously an implanted pulmonary artery pressure (IPAP) device is used for patients with severe limitations in their ability to perform physical activity who have been hospitalized for heart failure in the past year. The device consists of a battery-free sensor that is implanted permanently into the pulmonary artery: readings from the implant are transmitted to an external unit, which also wirelessly powers the implant as shown in Figure 8.12. The data is then transmitted to the patient's cardiologist. The system allows clinicians to stabilize pulmonary artery pressures by proactively managing medications and other treatment options, as well as providing an early indication of worsening heart failure.

The implanted component of the system is a miniature leadless pressure sensor, shown in Figure 8.12(b), placed via the femoral vein into the distal pulmonary artery under local anaesthesia. The pressure sensor is a few centimetres long and has a thin, curved nitinol wire at each end, which serves as an anchor in the pulmonary artery. The pressure sensor consists of an inductor coil and a pressure-sensitive capacitor encased in a hermetically sealed silica capsule covered by biocompatible silicone. The coil and capacitor form an electrical

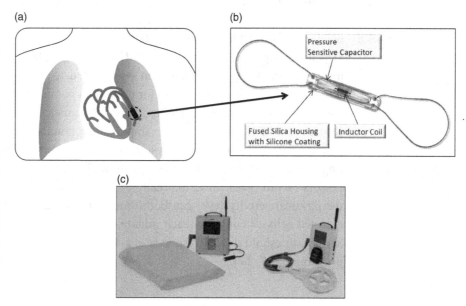

Figure 8.12

(a) A pressure sensor is implanted into the pulmonary artery. (b) The pressure sensor consists of an inductor coil and a pressure-sensitive capacitor, together forming a resonant circuit. (c) The readout electronics, which also can transmit the data via GSM to the physician, are placed in a pillow on which the patient lies. Figures reproduced with permission from St Jude Medical Inc.

circuit that resonates at a specific frequency as covered in Chapter 3. The capacitance depends on the pressure, as covered in Chapter 2, and any changes in pressure result in a change in capacitance and therefore a shift in the resonant frequency of the device. Electromagnetic coupling between the outside unit and the sensor uses an external antenna that is housed within a pillow on which the patient lies, as shown in Figure 8.15(c). When measurements are made, the transmitter is turned on and couples magnetically to the sensor, as described previously. The transmitting antenna measures the resonant frequency of the sensor circuit (see next paragraph), which is then converted to a pressure waveform. A pressure waveform is recorded for approximately 20 seconds and then transmitted. The device also has an atmospheric barometer, which automatically subtracts the ambient pressure from that measured by the implanted sensor. The system typically operates over a range of approximately 30.0 to 37.5 MHz, with a measurement bandwidth of 1 MHz.

Changes in resonant frequency of the sensor are determined by adjusting the phase and frequency of the transmitter until the transmitter frequency matches the resonant frequency of the sensor. The transmitter puts out a burst of RF energy with a predetermined frequency, or set of frequencies, and a predetermined amplitude. Magnetic coupling induces a current in the internal circuit and, in a reciprocal manner, the external detector picks up the magnetic field from this current, which occurs at a slightly different frequency from that transmitted. By using two phase-locked loops the transmitter frequency is adjusted until it matches exactly the received signal (in practice this is done by reducing the phase difference between the two signals), at which point the pressure can be derived.

8.4.1.5 Combining Data From Multiple Sensors

Depending upon which particular CIED is implanted, a large number of different cardiac measurements can be reported to the physician. Presenting this data in time-locked graphical form allows a complete picture of the current cardiac status to be formed, aiding in predictive measures to minimize unforeseen events in the future. For example, one commercial company produces graphs of up to ten different physiological metrics from one of their wireless implanted CIEDs: a subset of this data is shown in Figure 8.13. Additional sensors record data that is indirectly related to cardiac activity. For example, the physical activity level can be estimated by integrating the accelerometer-detected activity counts in each minute, with a minute counting as 'active' if above a threshold that corresponds to a walking rate of approximately 70 steps per minute [16, 17].

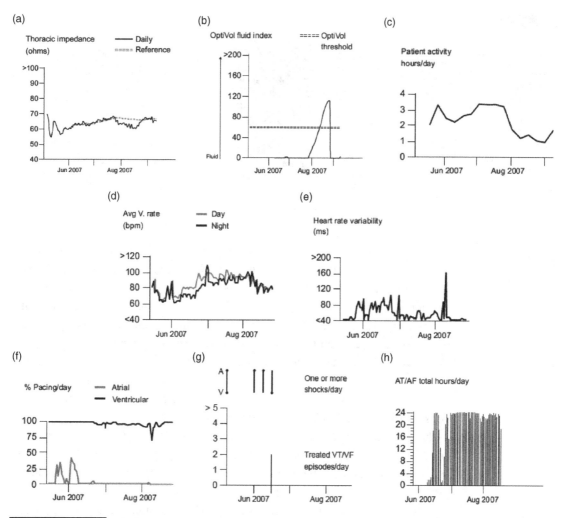

Figure 8.13 An illustration of the data readout from an implanted CIED that can be wirelessly transferred to a cardiologist. Based upon the trends in the particular measures a decision to hospitalize the patient can be made before serious symptoms occur. (a) Thoracic impedance and (b) derived fluid volume. (c) Patient activity from an accelerometer. (d) Average heart rate and (e) derived heart rate variability. (f) Amount of atrial and ventricular pacing produced by a DDD pacemaker, (g) the number of shocks applied per day to treat ventricular tachycardia or ventricular fibrillation, and (h) the total time spent in atrial tachycardia or atrial fibrillation on a daily basis.

8.4.1.6 New Technologies to Reduce the Risks of CIED Infection

As already stated, the number of patients annually who have a CIED implanted is increasing steadily. However, a major safety concern is that there is a significant infection rate (4 to 10%) among these patients, with a one-year mortality rate between 15 and 20% [18, 19]. The main infections are due to

Figure 8.14

(a) Photographs of two commercial leadless pacemakers, which can be implanted directly into the heart, as shown in (b), eliminating the need for a pacemaker electrode. Photographs reproduced with permission from St. Jude's Medical Inc. and Medtronic.

(a)

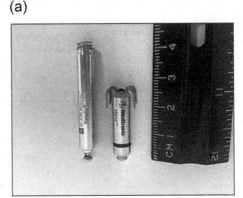

(b)

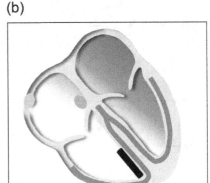

Staphylococcus aureus (*S. aureus*) and coagulase-negative staphylococcus species, which occur in or near the implantation pocket. There have been two main approaches to deal with this issue. The first is to use an antibacterial envelope in which to place the CIED, with the mesh envelope releasing antibacterial agents such as minocycline and rifampicin. The second approach is to reduce the complications associated with the requirement of having a transvenous electrode lead, and instead implanting a **leadless pacemaker** [20–23]. This device is small enough to be implanted in the heart itself, and so has no physical connection between the endocardium and the subcutaneous pocket [18]. These miniature devices, shown in Figure 8.14, can be delivered to the heart directly through a catheter inserted into the femoral vein. There are currently two different versions of leadless pacemakers on the market. There are several differences between the two devices but, from a design point of view, the most relevant is that rate response pacing is provided via a three-axis accelerometer in one design, and via a blood temperature sensor in the other.

Another approach has been the development and design of a completely subcutaneous ICD, which ensures that neither the heart nor the vasculature are impacted by the surgical implantation process. The generator for the ICD is placed subcutaneously over the ribcage. There are some limitations to this type of device compared to the fully implantable one described earlier, which means that it cannot be used for all patients. Specifically, the subcutaneous ICD cannot provide anti-tachycardia pacing for ventricular tachycardia and cannot perform measurements such as the impedance monitoring outlined earlier. Wearable cardioverter defibrillators have also been developed recently, consisting of external vests that automatically detect and treat tachyarrhythmias [24].

8.4.2 Continuous Glucose Monitors

As covered in Chapter 2 a patient with diabetes performs a 'finger test' up to three to four times a day to measure blood glucose levels. Based on the reading the dose and timing of the next insulin injection can be determined. There are a number of limitations, in terms of patient care, associated with being able to perform only a few measurements per day. For example, in the time between measurements the glucose level can rise above or fall below the desired range, and such changes are not detected. In addition, measurements of a few discrete time points several hours apart cannot be used to predict potential hypoglycaemic or hyperglycaemic episodes between measurements.

These issues, as well as the inconvenience of having to perform finger pricking, can be circumvented by using an implanted continuous glucose monitor (CGM). The device is surgically inserted ~5 mm under the skin, either in the abdomen or on the back of the upper arm, and measured glucose levels are transmitted wirelessly at a frequency of 433.6 MHz every few minutes to a nearby receiver. It is important to realize that the glucose level using a CGM is measured in the *subcutaneous interstitial fluid* (ISF) rather than in the blood as is the case for external devices. Glucose enters the ISF via rapid exchange from the blood and is taken up into the cells within this compartment. There is a time lag of between 5 and 15 minutes (average 6 to 7 minutes) for changes in glucose concentration to be detected in the ISF compared to in the blood [25]. As such CGM is intended to complement, not replace, information obtained from finger-prick values, and periodic calibration with respect to finger-prick blood measurements must be performed.

The principles of CGM are broadly similar to those covered in Chapter 2 for finger-stick measurements [26–28]. Figure 8.15 shows the design of a CGM, which consists of many planar layers stacked on top of one another. The set up is a three-electrode system, which produces a current proportional to the glucose concentration in the ISF. The three electrodes, working (carbon), reference (Ag/AgCl) and counter (carbon), are printed onto a polyester substrate with thin dielectric layers placed in between the electrodes. The chemistry involved is similar to that outlined in Chapter 2, with the core component of the CGM being a miniature amperometric glucose sensor. Since the device is implanted in the body, the enzyme and mediator must be immobilized onto the electrode surface, otherwise they would dissolve in the subcutaneous fluid and diffuse away from the electrode. GOx is immobilized on the working electrode, and an alternative mediator substituting for oxygen is also immobilized in a scheme referred to as a Wired Enzyme [28]. GOx produces hydrogen peroxide in an amount that is

Figure 8.15

Schematic showing the different components of a continuous glucose monitor. The outer membrane provides a glucose-limiting flux barrier and an inert biocompatible interface.

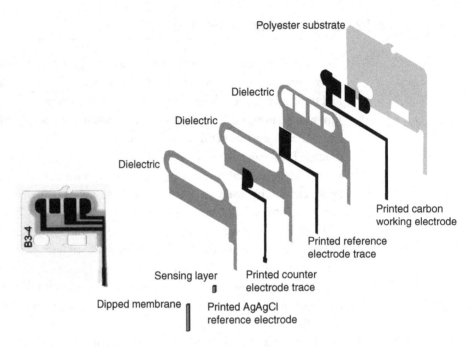

directly proportional to the concentration of glucose, and is measured amperometrically at a potential of approximately 0.7 volts using a platinum electrode.

A CGM implanted subcutaneously on the back of the upper arm is shown in Figure 8.16(a). The glucose level in mg/dL is updated once every minute and the value is transmitted every few minutes to an external unit, which displays a graph of glucose level over time, as shown in Figure 8.16(b). Audible alarms can be set for low or high glucose levels, or projected values based on values over time. Software associated with the system allows entry of events such as exercise, illness, calorific intake and insulin injection, as a way of tracking the causes of high or low glucose events.

The data from a CGM can enable a more patient-specific predictive model of glucose levels in response to, for example, extended exercise or carbohydrate consumption, which varies widely between different patients. Continuous glucose monitoring can also be combined with a small wearable insulin pump. Together these form a closed loop in which insulin can be injected in response to changes in glucose levels.

In addition to the electrochemical detection described above, there are several new technologies, including infrared absorption and other optical techniques, photoacoustic spectroscopy and transdermal sensors, which show promise for future generations of CGMs. In order to assess the performance of new types of glucose meter, one commonly used method is the Parkes error

Figure 8.16

(a) Electrode and wireless transmitter placed under the skin. (b) Display module for the continuous glucose monitoring unit. Figures reproduced with permission from Abbott Diabetes Care.

(a)

(b)

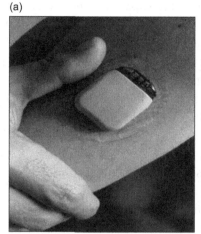
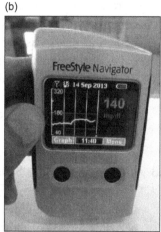

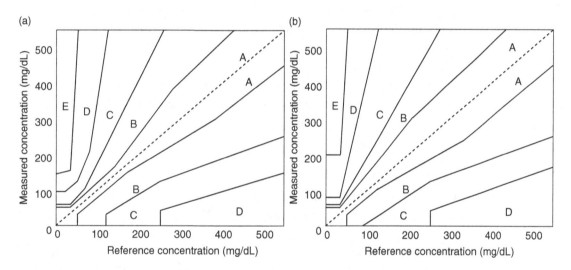

Figure 8.17 Parkes error graphs for glucose measurements: (a) type 1 diabetes; (b) type 2 diabetes.

grid [29], shown in Figure 8.17 for type 1 and type 2 diabetes patients. This grid is based on the responses of 100 scientists who were asked to assign risk to self-monitoring blood glucose systems. Values from the device are compared with a 'gold-standard' reference value from a validated laboratory measurement. Five zones (A to E) are defined, as shown in Figure 8.17. Zone A represents a clinically accurate measurement; zone B represents a measurement that is not accurate but benign in terms of its effect on the patient; zone C represents an inaccurate measurement (either too high or too low) that would produce an overcorrection in administered insulin dose; zone

D represents a failure to detect hypo- or hyperglycaemia; and zone E gives completely erroneous measurements. The general 'rule of thumb' based on the definition of the risk boundaries is that a new meter for clinical use should have at least 95% of its data points in zone A of the Parkes grid. Some manufacturers have elected to apply the Parkes metric to their products by specifying that data points in both the A and B zones are clinically acceptable; however, most clinicians currently do not support that claim.

8.4.3 Implanted Pressure Sensors for Glaucoma

Glaucoma is a condition that results in a gradual decrease in the visual field and accounts for 8% of global blindness [30]. Once lost, vision cannot normally be recovered, so once diagnosed treatment is aimed at preventing further vision loss [31]. For 50% of glaucoma patients the symptoms are caused by an increase in ocular pressure [32]. If this condition can be detected early enough, it is possible to slow the progression of the disease by pharmaceutical and surgical means, both of which reduce the IOP.

Monitoring the progression of glaucoma while the patient is being treated involves taking regular measurements of the patient's IOP. This is typically done at the hospital with readings taken weeks or months apart, a problem since the IOP can rise and fall unpredictably throughout the day. An implanted continuous monitoring device would therefore be of very practical importance. Figure 8.18 shows one such device. The ring-shaped IOP sensor is a device with eight pressure-sensitive capacitors combined with a circular microcoil antenna, all integrated with a single ASIC [33]. The resonant frequency of the device depends upon the capacitance, which is a function of the IOP. The device is completely encapsulated in biocompatible silicone, with an outside diameter of 11.2 mm and a thickness of 0.9 mm. The implant is inductively coupled to an external transmitter/receiver, operating at 13.56 MHz, which both powers the device and also transfers the data. The external device is battery powered and during IOP measurements, which last a few seconds (10 to 20 signal averages), the reader unit is held within 5 cm of the front of the eye.

Although this device is not yet on the market, it is being tested on patients who have undergone cataract surgery in which a polymeric intraocular lens (IOL) has been implanted. Initial results after one year have shown good patient tolerance, and readings with a high degree of accuracy with respect to conventional external methods [34].

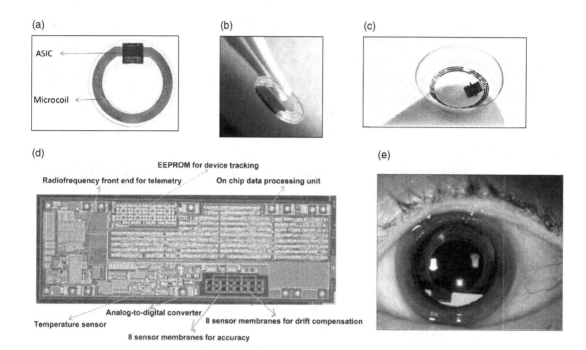

Figure 8.18   Photographs of a MEMS-based ocular sensor to measure intraocular pressure in glaucoma patients.
(a) Microcoil used to receive power from an external source, connected to an application-specific integrated circuit. (b) and (c) Photographs of the integrated sensor. (d) Photograph of the integrated electronics including the eight capacitive pressure sensors (as well as a thermal sensor). (e) Photograph of the device implanted in a patient. Figures reproduced with permission from Todani *et al.* [33].

8.5 Packaging for Implanted Devices

Devices that are temporarily or permanently implanted in the body must be made as inert as possible with respect to the body's immune system and must also not interact with the body causing the device to malfunction [35]. The body represents a high humidity and conductivity environment at 37°C, as well as containing a variety of reactive chemical species such as proteins, enzymes and cellular structures. Immunological responses are a major obstacle to successful implantation, particularly if the sensor element itself has to come into close contact with the body. Adhesion of proteins can change the physical characteristics of the device, and alter its performance, particularly if the degree of adhesion changes over time. For example, corrosion can occur in intravascular sensors, and macrophage aggregation often occurs with subcutaneous or intramuscular implantation. Many sensors, e.g. pacemakers and other CIEDs, are protected and packaged within relatively large metallic cases, but such bulky solutions are not available for much smaller sensors. Several chemical groups such as polyethylene glycol have been proposed as being 'protein resistant' [36], and silicon-based polymers are often used to provide an inert barrier to sensors. Guidelines have been developed for new biocompatible materials, ISO 10993–1 being one such example. More details on biocompatibility and safety tests are covered in section 9.6.

Appendix: Reference Standards and Information Related to Wireless Implant Technology

ANSI/AAMI PC69:2007, Active implantable medical devices – Electromagnetic compatibility – EMC test protocols for implantable cardiac pacemakers and implantable cardioverter defibrillators

ANSI C63.4:2009, American National Standard for Methods of Measurement of Radio-Noise Emissions from Low-Voltage Electrical and Electronic Equipment in the Range of 9 kHz to 40 GHz

ANSI C63.18:1997, American National Standard Recommended Practice for an On-Site, Ad Hoc Test Method for Estimating Radiated Electromagnetic Immunity of Medical Devices to Specific Radio-Frequency Transmitters

IEC 60601-1-2:2007, Medical Electrical Equipment – Part 1–2: General Requirements for Basic Safety and Essential Performance – Collateral Standard: Electromagnetic Compatibility – Requirements and Tests

IEEE P11073-00101™-2008 – Guide for Health Informatics – Point-of-Care Medical Device Communication – Guidelines for the Use of RF Wireless Technology. There are several standards under the IEEE 11073 family that address health informatics point-of-care medical device communications and provide useful information

Most ISO standards for medical electrical equipment reference clauses in IEC 60601–1, including Clause 17 (previously Clause 36) and IEC 60601–1-2

ISO/TR 16056–1, Health Informatics – Interoperability of Telehealth Systems and Networks – Part 1: Introduction and Definitions. Part 2: Real-time Systems

ISO 14708–1, Implants for Surgery – Active Implantable Medical Devices – Part 1: General Requirements for Safety, Marking, and for Information to be Provided by the Manufacturer. Part 2: Cardiac Pacemakers

ISO 14971 Second Edition 2007–03-01 Medical Devices – Application of Risk Management to Medical Devices

ISO 14117, Active Implantable Medical Devices – Electromagnetic Compatibility – EMC Test Protocols for Implantable Cardiac Pacemakers, Implantable Cardioverter Defibrillators, and Cardiac Resynchronization Devices

PROBLEMS

8.1 Research different devices that people have used to create a humidity sensor to measure sweat rate. Include the detection principle and the basic design of the device. Estimate the sensitivity of the device.

8.2 Devise a sensor that can be integrated into a T-shirt to measure surface temperature. What are the potential interference signals for this device? Estimate the sensitivity of the device.

8.3 Design some form of energy harvesting system using the body's thermal properties to power a sensor.

8.4 Explain how the full diode bridge rectifying circuit shown in Figure 8.7(a) works.

8.5 Derive the frequency-dependent impedance of the mutually coupled circuits shown in Figure 8.7(b). Show the simplification that occurs at resonance.

8.6 Design a system for transmitting data through the body at 403 MHz. Include realistic inductor values for a helical antenna (you will need to look up formulae for calculating the inductance of a solenoid).

8.7 If the coupling constant in a mutually coupled wireless configuration is 0.03, and the Q values are $Q_1 = 1000$, $Q_2 = 10$, calculate the maximum power efficiency. How is this affected if Q_1 increases to 10 000? How about if $Q_1 = 1000$ and $Q_2 = 100$? Explain how your results define the design criteria for the individual circuits.

8.8 Design a system for a patient to generate internal power based on the Seebeck effect (see section 8.3.1). Pay attention to the sensors needed, the placement of these sensors and the amount of power that can be generated.

8.9 (a) Suggest reasons why ASK has a lower immunity to noise than FSK or PSK? (b) Why does FSK require a higher bandwidth than either ASK or PSK?

8.10 Draw the signal that is transmitted through the body for a digital input of 10010101 using: (a) ASK modulation, (b) FSK modulation and (c) PSK modulation.

8.11 Future developments in cardiac sensors may include the measurement of pH. Suggest mechanisms by which this could be performed, and sketch out possible devices.

8.12 Design a combined capacitor/inductor sensor that: (i) increases in resonance frequency with increased pressure, (ii) decreases in resonance frequency with increased pressure.

8.13 Many sensors work by detecting a change in resonance frequency of an implanted device. In addition to the methods outlined in this chapter, design a new internal/external circuit for detecting the changes.

8.14 Design a new health app using the sensors in your current smartphone.

8.15 Provide three designs for a step-counting algorithm using a smartphone.

8.16 By investigating the antenna literature, design a PIFA that can be used to transmit energy through the body at 440 MHz.

8.17 What happens to an ultra-wideband signal as it passes through tissue in terms of its amplitude and frequency characteristics?

REFERENCES

[1] Barello, S., Triberti, S., Graffigna, G. *et al.* eHealth for patient engagement: a systematic review. *Front Psychol* 2015; **6**:2013.

[2] Graffigna, G., Barello, S., Bonanomi, A. & Menichetti, J. The motivating function of healthcare professional in ehealth and mhealth interventions for type 2 diabetes patients and the mediating role of patient engagement. *J Diabetes Res* 2016.

[3] Tang, P. C., Overhage, J. M., Chan, A. S. *et al.* Online disease management of diabetes: Engaging and Motivating Patients Online with Enhanced Resources-Diabetes (EMPOWER-D), a randomized controlled trial. *J Am Med Inform Assn* 2013; **20**(3):526–34.

[4] Zheng, Y. L., Ding, X. R., Poon, C. C. *et al.* Unobtrusive sensing and wearable devices for health informatics. *IEEE Trans Biomed Eng* 2014; **61**(5):1538–54.

[5] Vilela, D., Romeo, A. & Sanchez, S. Flexible sensors for biomedical technology. *Lab Chip* 2016; **16**(3):402–8.

[6] Bandodkar, A. J., Hung, V. W. S., Jia, W. Z. *et al.* Tattoo-based potentiometric ion-selective sensors for epidermal pH monitoring. *Analyst* 2013; **138**(1):123–8.

[7] Bandodkar, A. J., Jia, W. Z. & Wang, J. Tattoo-based wearable electrochemical devices: a review. *Electroanal* 2015; **27**(3):562–72.

[8] Bandodkar, A. J., Jia, W. Z., Yardimci, C. *et al.* Tattoo-based noninvasive glucose monitoring: a proof-of-concept study. *Anal Chem* 2015;**87**(1):394–8.

[9] Jia, W. Z., Bandodkar, A. J., Valdes-Ramirez, G. *et al.* Electrochemical tattoo biosensors for real-time noninvasive lactate monitoring in human perspiration. *Anal Chem* 2013; **85**(14):6553–60.

[10] Kiourti, A., Psathas, K. A. & Nikita, K. S. Implantable and ingestible medical devices with wireless telemetry functionalities: a review of current status and challenges. *Bioelectromagnetics* 2014; **35**(1):1–15.

[11] Kiourti, A. & Nikita, K. S. Miniature scalp-implantable antennas for telemetry in the MICS and ISM bands: design, safety considerations and link budget analysis. *IEEE T Antenn Propag* 2012; **60**(8):3568–75.

[12] Hannan, M. A., Abbas, S. M., Samad, S. A. & Hussain, A. Modulation techniques for biomedical implanted devices and their challenges. *Sensors* 2012; **12**(1):297–319.

[13] Abraham, W. T., Stevenson, L. W., Bourge, R. C. *et al.* Sustained efficacy of pulmonary artery pressure to guide adjustment of chronic heart failure therapy: complete follow-up results from the CHAMPION randomised trial. *Lancet* 2016; **387**(10017):453–61.

[14] Sanders, R. S. & Lee, M. T. Implantable pacemakers. *Proc IEEE* 1996; **84**(3): 480–6.

[15] Yu, C. M., Wang, L., Chau, E. *et al.* Intrathoracic impedance monitoring in patients with heart failure: correlation with fluid status and feasibility of early warning preceding hospitalization. *Circulation* 2005; **112**(6):841–8.

[16] Merchant, F. M., Dec, G. W. & Singh, J. P. Implantable sensors for heart failure. *Circ Arrhythm Electrophysiol* 2010; **3**(6):657–67.

[17] Adamson, P. B., Smith, A. L., Abraham, W. T. *et al.* Continuous autonomic assessment in patients with symptomatic heart failure: prognostic value of heart rate variability measured by an implanted cardiac resynchronization device. *Circulation* 2004; **110**(16):2389–94.

[18] Kondo, Y., Ueda, M., Kobayashi, Y. & Schwab, J. O. New horizon for infection prevention technology and implantable device. *J Arrhythm* 2016; **32**(4):297–302.

[19] Baman, T. S., Gupta, S. K., Valle, J. A. & Yamada, E. Risk factors for mortality in patients with cardiac device-related infection. *Circ Arrhythm Electrophysiol* 2009; **2**(2):129–34.

[20] Spickler, J. W., Rasor, N. S., Kezdi, P. *et al.* Totally self-contained intracardiac pacemaker. *J Electrocardiol* 1970; **3**(3–4):325–31.

[21] Knops, R. E., Tjong, F. V., Neuzil, P. *et al.* Chronic performance of a leadless cardiac pacemaker: 1-year follow-up of the LEADLESS trial. *J Am Coll Cardiol* 2015; **65**(15):1497–504.

[22] Koruth, J. S., Rippy, M. K., Khairkhahan, A. *et al.* Feasibility and efficacy of percutaneously delivered leadless cardiac pacing in an in vivo ovine model. *J Cardiovasc Electrophysiol* 2015; **26**(3):322–8.

[23] Reddy, V. Y., Knops, R. E., Sperzel, J. *et al.* Permanent leadless cardiac pacing: results of the LEADLESS trial. *Circulation* 2014; **129**(14):1466–71.

[24] Kondo, Y., Linhart, M., Andrie, R. P. & Schwab, J. O. Usefulness of the wearable cardioverter defibrillator in patients in the early post-myocardial infarction phase with high risk of sudden cardiac death: a single-center European experience. *J Arrhythm* 2015; **31**(5):293–5.

[25] Oliver, N. S., Toumazou, C., Cass, A. E. & Johnston, D. G. Glucose sensors: a review of current and emerging technology. *Diabet Med* 2009; **26**(3):197–210.

[26] McGarraugh, G. The chemistry of commercial continuous glucose monitors. *Diabetes Technol Ther* 2009; **11** Suppl 1:S17–24.

[27] Khadilkar, K. S., Bandgar, T., Shivane, V., Lila, A. & Shah, N. Current concepts in blood glucose monitoring. *Indian J Endocrinol Metab* 2013; **17**(Suppl 3): S643–9.

[28] Heller, A. & Feldman, B. Electrochemistry in diabetes management. *Acc Chem Res* 2010; **43**(7):963–73.

[29] Pfutzner, A., Klonoff, D. C., Pardo, S. & Parkes, J. L. Technical aspects of the Parkes error grid. *J Diabetes Sci Technol* 2013; **7**(5):1275–81.

[30] King, A., Azuara-Blanco, A. & Tuulonen, A. Glaucoma. *BMJ* 2013; 346.

[31] Weinreb, R. N. Ocular hypertension: defining risks and clinical options. *Am J Ophthalmol* 2004; **138**(3):S1–2.

[32] Sommer, A., Tielsch, J. M., Katz, J. *et al.* Relationship between intraocular-pressure and primary open angle glaucoma among White and Black-Americans: the Baltimore Eye Survey. *Arch Ophthalmol* 1991; **109**(8): 1090–5.

[33] Todani, A., Behlau, I., Fava, M. A. *et al.* Intraocular pressure measurement by radio wave telemetry. *Invest Ophth Vis Sci* 2011; **52**(13):9573–80.

[34] Koutsonas, A., Walter, P., Roessler, G. & Plange, N. Implantation of a novel telemetric intraocular pressure sensor in patients with glaucoma (ARGOS study): 1-year results. *Invest Ophth Vis Sci* 2015; **56**(2):1063–9.

[35] Clausen, I. & Glott, T. Development of clinically relevant implantable pressure sensors: perspectives and challenges. *Sensors* 2014; **14**(9):17686–702.

[36] Kingshott, P. & Griesser, H. J. Surfaces that resist bioadhesion. *Curr Opin Solid St Mater Sci* 1999; **4**(4):403–12.

9 Safety of Biomedical Instruments and Devices

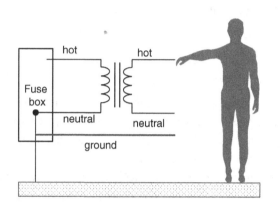

Introduction

Every year many thousands of patients receive some form of electrical shock from medical devices and instruments. These are typically not fatal and do not cause serious injury, but rather result in significant discomfort to the patient and may cause effects such as ventricular fibrillation. There are many incidents of physicians and nurses also receiving electric shocks. These events occur despite strict regulations to ensure the **electrical safety** of all medical devices. Although some events occur due to a design flaw in a particular device, the most common situation in which electrical shocks occur is one in which a patient is hooked up to many different devices in the operating room (OR), as shown in Figure 9.1. These devices might include defibrillators, pacemakers, ultrasound scanning equipment, cardiac catheters and EEG equipment to monitor anaesthesia. In addition, there are a number of ancillary devices such as high intensity lights and intravenous lines, which are attached to, or close to, the patient. Conductive body fluids also result in unintended conduction paths close to the patient or in the medical device itself if these fluids come into contact with the equipment. In general, medical equipment, and particularly the power cords and power cables feeding the equipment, see many years of use, and it is estimated that over 70% of electrical 'incidents' occur due to wear and tear on the power cords, which are regularly run over or mechanically stressed. The electrical insulation around equipment also becomes damaged over time during normal operation. In order to govern the design of medical equipment, regulatory bodies have produced documents to safeguard patients, and the IEC has produced a standard, the IEC 60601 Medical Electrical Equipment – Part 1: General Requirements for Safety and Essential Performance. A brief description of the tests that must be performed before installation, and for regular maintenance, is covered in section 9.5 of this chapter.

Figure 9.1

Multiple pieces of electrical equipment may be used together in an operating room.

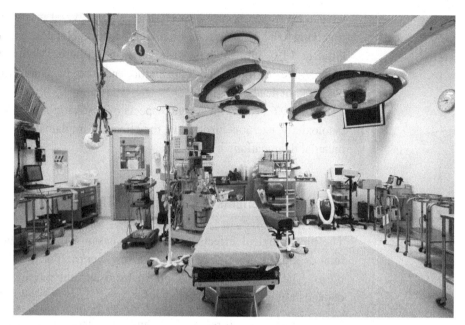

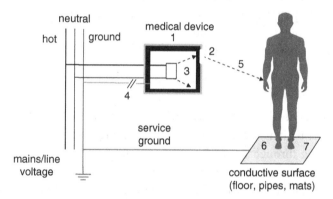

Figure 9.2   Points of electrical safety analysis for medical devices. (1) A part of line/mains-powered equipment is within reach of the patient. (2) Metal parts are exposed. (3) Damage to the equipment has resulted in the enclosure/chassis becoming electrified. (4) Exposed metal parts are ungrounded. (5) The patient or physician makes contact with an exposed (potentially live) surface. (6) A second grounded conductive surface is in contact with the medical device. (7) The patient or physician makes contact with this second grounded surface.

One can analyze the different potential danger points in terms of the possibility that multiples of these are present, and can cause a potential electrical shock to the patient or physician, as shown in Figure 9.2.

The other side of safety in medical devices and instrumentation is biocompatibility. As described in Chapter 1, despite all of the legal safety requirements in place, it is estimated that roughly 1.5% of 501(k) predicate devices have to be recalled every year. There are also several instances of major recalls involving tens or hundreds of patient deaths. Notable examples include the Guidant LLC implantable cardioverter–debrillator, Medtronic Sprint Fidelis implantable cardioverter–defibrillator and the St. Jude Medical Riata ST defibrillator leads.

Section 9.6 deals with several aspects of assessing biocompatibility including cytotoxicity, haemocompatibility, the effects of implantation, carcinogenicity and material degradation.

9.1   Physiological Effects of Current Flow Through the Human Body

The physiological effects of a current passing through the body to ground include muscle cramps, respiratory arrest, ventricular fibrillation and physical burns. The degree of danger posed by a current passing through the body depends upon:

(i) the amount of current and the duration for which the current flows
(ii) the pathway of the current through the body (essentially what fraction passes through the heart), and
(iii) whether the current is DC or AC.

For example, the heart can be stopped by a current of 40 mA applied for 250 ms, or one of 200 mA applied for 50 ms (a current passing through the chest can cause ventricular fibrillation or asphyxia due to tetany of the respiratory muscles).

The main natural protection against a high current (from an external source) passing through the body is the very high resistance of the skin, as shown in Figure 9.3(a). The electrical resistance of the skin at 50/60 Hz is between 15 kΩ and 1 MΩ/cm^2 depending on which part of the body is being considered. However, if the skin becomes wet (for example when an ECG or EEG electrode using a conducting gel is used) or there is a wound that has broken the skin, then the resistance can be reduced to as little as 100 to 1000 Ω. Intravenous catheters also reduce the effective resistance of the body, since the fluid has a much lower resistance than the skin. If a person now makes contact with a live wire at 240/110 volts, the current passing through the body is of the order of amps, as opposed to micro- to milliamps if the skin were dry. The current passes through the entire body before entering the earth, as shown in Figure 9.3(b). This current can cause muscles to contract meaning that it is very difficult for the subject to detach themselves from the live source. This type of electrical shock is called a **macroshock**, since the current passing through the body is very high.

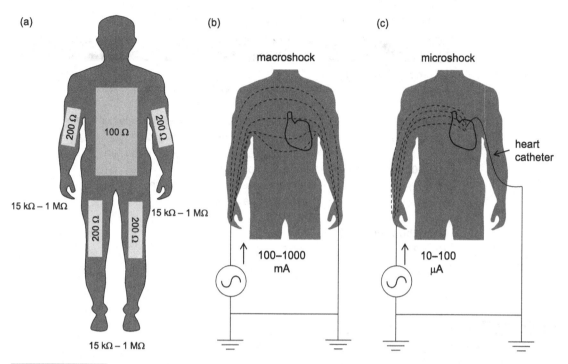

Figure 9.3 (a) Approximate resistance values of different parts of the body. (b) Current paths (dotted lines) producing an electric shock from a macrocurrent (>100 mA), called a macroshock. (c) Corresponding current paths from a microcurrent (<1 mA), termed a microshock. Note the concentration of current through the heart due to the presence of a catheter in the latter case.

However, it does not require very large currents to produce a dangerous electrical shock to the patient. **Microshock** refers to the effect of a much smaller current, as low as hundreds of micro-amps, which is introduced directly into the body via a highly conducting avenue, such as a pacemaker wire, or a fluid-filled device, such as a central venous pressure catheter, as shown in Figure 9.3(c). As will be seen later in this chapter, the low current levels associated with microshocks can also be produced by leakage currents, which can occur if the equipment is at a different potential than the ground.

Table 9.1 lists the physiological effects of macroshock and microshock currents experienced for 1 second.

One of the problems concerning electric shock and patients is that they may not be able to feel electric current passing through their bodies due to being anaesthetized, sedated or on pain medication. They may also have a lower let-go limit than normal, i.e. even if they do feel the current, if the current is above a certain threshold level, they will be unable to 'let-go' due to muscle contractions. Figure 9.4 shows the frequency dependence of the sensation threshold and

Table 9.1 Effects of current flow through the body for 1 second

	Effect
Macroshock current	
1 mA	Perception threshold
5 mA	Maximum 'harmless' current
10–20 mA	Let-go current before sustained muscle contraction
50 mA	Mechanical injury and pain
	Heart and respiratory functions maintained
100–300 mA	Ventricular fibrillation
>1 A	Sustained myocardial contraction
	Respiratory paralysis
	Tissue burns
Microshock current	
100 µA	Ventricular fibrillation

Figure 9.4

The sensation threshold and let-go limit currents as a function of frequency.

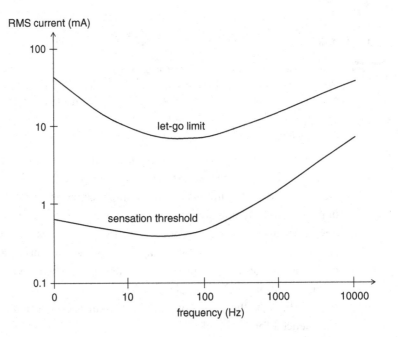

let-go limit for the general population. One notable feature is that the minimum let-go current thresholds occur approximately at 50/60 Hz, exactly the frequency range where most of the problems occur due to coupling to the mains power supply!

9.2 The Hospital Electrical Supply

The electrical demands of a hospital are far greater than those for other types of building of similar size. One reason is that hospitals house equipment such as magnetic resonance imaging and computed tomography scanners with very high power requirements. Air conditioning and ventilation systems can be two or three times larger due to the increased air exchange rate necessary, more stringent temperature requirements and higher equipment loads. Many specialized areas in a hospital have a large number of receptacles since many devices may be used at one time. For example, a hospital operating room of 50 m^2 may include 30 or more electrical receptacles, many more than the equivalent office space would need. Hospital lighting typically has very high lumen and power requirements based upon use in surgical, examination and medical procedure rooms. In addition, the hospital's emergency system is much larger than that of other building types. Hospitals require emergency power to sustain patients who are attached to monitoring or life-sustaining equipment, and are therefore not able to evacuate in the event of a power outage (note that in many countries electrical outlets on the emergency supply are a different colour, for example red in the United States). Emergency power is effectively needed indefinitely, not just long enough to safely exit the building as is the case elsewhere. A typical 20 000 m^2 hospital needs ~2500 kW or more of generator capacity, and the emergency power requirements of a hospital may exceed 50 or 60% of its total power needs.

A simplified diagram of a hospital electrical supply is shown in Figure 9.5. The power coming into the building is at 4800 volts at 50/60 Hz: this is stepped down to 240 volts inside the building using a transformer. Depending upon geographical location, this supply can either be centre-tapped down to two voltages of 120 volts, or can be used as a single 240 volt supply.

One of the most important considerations when multiple instruments are all powered on is that all of the grounds are actually at the same potential. Figure 9.5 shows a protective earth bar and an equipotential grounding bar as an integral part of the electrical system. As noted earlier, even a very small current can cause microshock in a patient, and this can easily arise if different pieces of equipment are at slightly different potentials. Examples of equipotential bonding bars are shown in Figure 9.6.

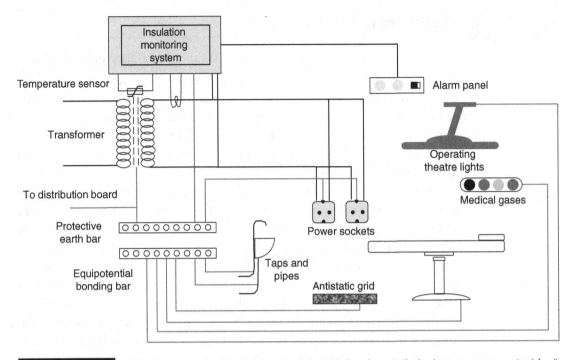

Figure 9.5   Simplified schematic of the hospital electrical supply (not shown is the back-up emergency system) feeding the individual outlets in a patient room.

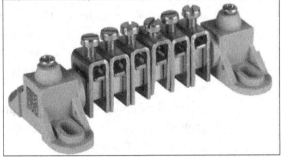

Figure 9.6   Two examples of setups for equipotential bonding.

9.2.1 Hospital-Grade Receptacles

The National Electric Code (NEC) Article 517–13 in the United States requires that all receptacles be 'hospital-grade'. This term means that the receptacle must be grounded by a separate insulated copper conductor, as well as having much higher overall mechanical strength and durability than normal household receptacles. Similar standards exist in many other countries. There are also corresponding hospital-grade power cords. Hospital-grade receptacles are required only in patient care areas such as operating rooms, ICU rooms, patient rooms, emergency department examination rooms and labour/delivery rooms. The primary reason to use a hospital-grade receptacle is that a receptacle with a greater contact tension minimizes the possibility that a plug supplying medical or life-support equipment might slip out of the receptacle due to unwanted tension, for example becoming entangled with another piece of equipment. Another important consideration in whether a hospital-grade receptacle is required is the type of equipment that will be connected to the receptacle. Some types of medical equipment include a manufacturer marking that indicates that proper grounding can only be ensured where connected to a hospital-grade receptacle.

In Europe, the electrical plug is the CEE 7/4 and is referred to as the 'Schuko plug', which derives from 'Schutzkontakt', meaning protection contact in German. The CEE 7/4 plug can be inserted in either direction into the receptacle since the hot and neutral lines are interchangeable. There are two grounding clips on the side of the plug. In normal operation current flows into a device from the hot terminal, and the neutral terminal acts as the current return path. The ground wire has no current flowing through it. Photographs of hospital-grade receptacles in the United States and continental Europe are shown in Figure 9.7.

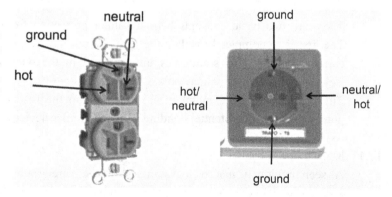

Figure 9.7

(left) Hospital-grade receptacles used in the United States. (right) European equivalent with interchangeable hot/neutral terminals.

9.3   Macroshock, Microshock and Leakage Currents: Causes and Prevention

As outlined earlier the effects of current passing through the body can be broken down into two broad classes, macroshock and microshock, where the difference refers to the magnitude of the current. Macroshock refers to currents on the hundreds of milli-amps, which can be caused by a patient touching a device at a very high potential. Microshock refers to a much smaller current, of the order of hundreds of micro-amps, which is introduced directly into the body via a highly conducting avenue such as a pacemaker wire or a fluid-filled device such as a central venous pressure catheter. Figure 9.8 shows sources for macroshock and microshock currents, as well as an additional pathway for general leakage currents.

9.3.1   Macroshock

As shown in Figure 9.8(a) macroshock can occur when there is an unintended connection between the hot terminal and the chassis due to a broken connection inside the device. This places the chassis at a very large potential. If the ground wire is properly connected to the chassis, then there is a large current flow through this wire. In this case, if the patient touches the chassis then only a relatively small current flows through the patient (since the body's resistance is much larger than that of the grounding wire). The high current flow to ground then trips the circuit breaker. However, if the ground wire is broken, then a large current can flow to ground through the patient, and a macroshock can occur.

9.3.2   Protection Against Macroshock

There are two basic concepts behind patient protection from electrical shock. The first is to completely isolate the patient from any grounded surface and all current sources. This strategy is implemented in the operating theatre using isolation transformers, which isolate the power output from the electrical ground. The second concept is to maintain all surfaces within the patient's reach at the same potential using equipotential bonding bars and similar devices.

9.3.2.1   Isolated Power Systems

As seen previously, macroshock can occur if there are ground faults, for example a short circuit occurs between the hot terminal and the ground conductor, which results in large currents flowing in the ground conductor (before the circuit

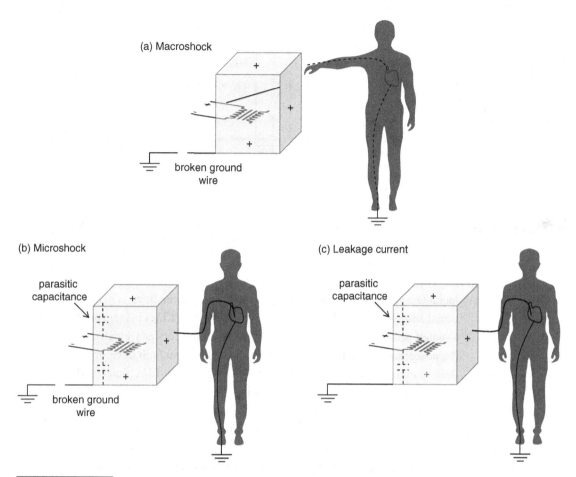

Figure 9.8   (a) An example of a condition that can give rise to a macroshock. A ground fault causes the medical instrument chassis to become live, i.e. to have a high potential compared to ground. If the ground wire is intact then the majority of the current flows to ground. If, however, the ground wire is broken then all of the current passes through the patient to ground. (b) An example of microshock. Even if there is no ground fault, the chassis can have a higher potential than ground due to capacitive coupling to the power cable. If there is a connection between the device and the patient, and the ground wire is faulty, then a current of several tens of micro-amps can flow through the patient. (c) The term leakage current refers to any current passing through the patient even when the ground wire is intact. If there is a direct connection to the heart then a significant fraction of the leakage current passes through the patient.

breaker trips). This results in high potential differences between different conductive surfaces. This type of ground fault can be prevented by using an isolated power system, usually implemented using an isolation transformer, as shown in Figure 9.9.

If there is perfect isolation, then the patient is completely protected from any shocks. However, as mentioned earlier, there is always some degree of electrical

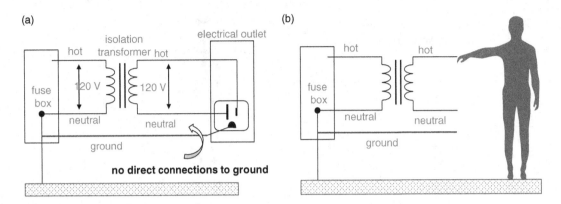

Figure 9.9 (a) An isolated and ungrounded power supply. (b) Even if a patient touches the hot wire or a medical device at high potential, there is no path for the current to flow through the body to ground and therefore the patient is safe.

leakage and coupling between equipment and power lines. Therefore an additional safety device called a line isolation monitor (LIM) is incorporated into an isolated power system. The LIM continually monitors any electrical leakage in the power supply system. For example, if an electrical fault develops in a medical device connected to an isolated power outlet, the LIM detects the leakage current, and sets off an alarm indicating the level of leakage current, but does not shut off the electric supply. The LIM also monitors how much power is being used by the equipment connected to it. If this is too much, the LIM will sound an alarm and indicate that there is an overload. Failure to reduce the load on the LIM will result in the circuit breaker tripping and loss of power to the circuit.

A schematic of a LIM is shown in Figure 9.10. The LIM works by having a switch that produces an artificial ground fault in one of the supply lines. If the other line has an actual ground fault, then a large current flows to ground and produces an alarm. As outlined above, the LIM can also monitor low leakage currents from either of the two isolated lines to ground: these currents can arise from parasitic capacitance, or compromised insulation between the isolated lines and the ground. The LIM can be incorporated into the operating room circuit breaker box or on the wall.

9.3.2.2 Ground Fault Circuit Interrupters

The second method to protect against macroshock is to incorporate a ground fault circuit interrupter (GFCI), which works by comparing the current flowing in the hot and neutral wires of a grounded power system. If a macroshock event occurs, then the current in the hot wire temporarily exceeds that in the neutral wire. The GFCI detects this and shuts off power to the system. The threshold for the

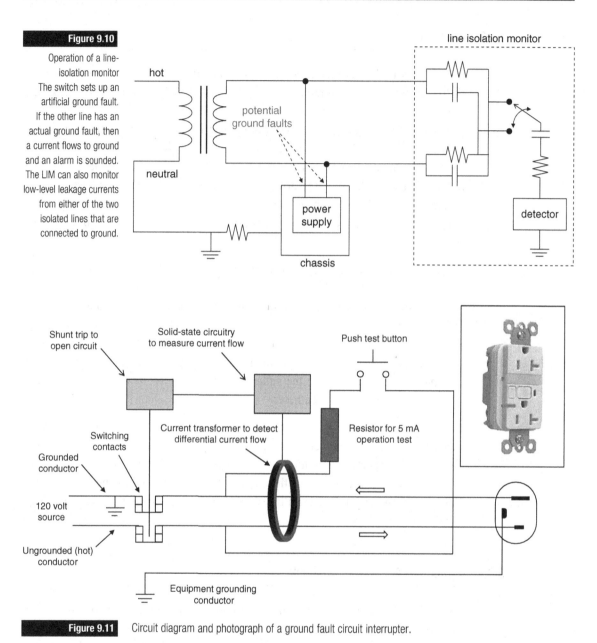

Figure 9.10

Operation of a line-isolation monitor The switch sets up an artificial ground fault. If the other line has an actual ground fault, then a current flows to ground and an alarm is sounded. The LIM can also monitor low-level leakage currents from either of the two isolated lines that are connected to ground.

Figure 9.11   Circuit diagram and photograph of a ground fault circuit interrupter.

imbalance between currents in the hot and neutral wires is 5 mA for most CGCIs. Figure 9.11 shows a circuit diagram of a GFCI. The disadvantage of using a GFCI in a critical care situation is the danger of loss of power to life-support equipment, and so these are typically not used in such situations.

9.3.3 Microshock

Microshock is characterized as an electric shock that can be produced by currents as low as 100 µA (below the perception threshold), which pass directly through the heart and can cause ventricular fibrillation. This magnitude of current can be produced, for example, by a small potential difference induced on the medical instrument chassis via the parasitic capacitance between the hot line and the chassis. Compared to the macroshock case there is a much smaller current passing through the patient; however, as outlined above, if this were to be conducted directly through the heart via a pacemaker wire, for example, then ventricular fibrillation could occur. Microshocks usually arise from one of three sources:

(i) a cardiac ECG lead acts as a current source and transfers the current to a catheter

(ii) a cardiac ECG lead acts as a current sink and current flows to a pacemaker lead, and

(iii) a ground loop occurs when a patient is connected to two grounded devices, one via a pacemaker lead, and then a third faulty device (not connected to the patient) is connected to the same power outlet circuit, which sends a high leakage current to the catheter.

Leakage currents occur even if there is no fault in the system. The two dominant mechanisms for leakage currents are capacitance associated with AC power systems and imperfect insulation between the load and surrounding metal work, which creates an alternative pathway for current to return to the neutral line. The maximum current leakage allowable through electrodes or catheters contacting the heart is 10 µA.

9.3.4 Protection Against Microshock

As outlined previously, the main danger from microshock occurs when devices are directly connected to the heart. Manufacturers provide the maximum security against microshock by using high levels of insulation: for example pacemaker wires and modules are very well electrically insulated. In terms of the electrical power supply, residual current devices (RCDs) are used in patient treatment areas to sense leakage currents flowing to earth from the equipment. If a significant leakage current flows, the RCD detects it and shuts off the power supplied to the equipment within 40 ms. A hospital RCD will trip at 10 mA leakage current (much lower than the >30 mA for household RCDs). Power outlets supplied through an RCD have a 'supply available' lamp. The lamp will extinguish when the RCD trips due to excessive leakage current.

Leakage currents can be minimized by using special low-leakage (<1 µA) power cords, which use insulation materials that reduce the capacitance between the live wires and the chassis. Equipotential earthing is installed in rooms classified as 'cardiac protected' electrical areas, and is intended to minimize any voltage differences between earthed parts of equipment and any other exposed metal in the room. This reduces the possibility of leakage currents that can cause micro-electrocution when the patient comes into contact with multiple items of equipment, or if the patient happens to come into contact with metal items in the room while they are connected to a medical device. All conductive metal in an equipotential area is connected to a common equipotential earth point with special heavy-duty cable.

9.4   Classification of Medical Devices

Body protected areas are designed for procedures in which patients are connected to equipment that lowers the natural resistance of the skin. So-called 'applied parts' such as electrode gels, conductive fluids entering the patient, metal needles and catheters provide a pathway for current to flow. RCDs, isolation transformers and LIMs are used in body-protected areas to reduce the chances of electrical shock from high leakage currents.

Cardiac protected areas are those in which the operating procedure involves placing an electrical conductor within or near the heart. In these areas protection against fibrillation due to small leakage currents is required. Electrical conductors used in these procedures include cardiac pacing electrodes, intracardiac ECG electrodes and intracardiac catheters. Equipotential grounding together with RCDs or LIMs is used to provide protection against microshock in cardiac-related procedures.

All devices in a hospital environment are classified according to class and type, depending upon their use and level of safety required. Stickers must be attached to each medical device to outline its class and type. Examples are shown in Figure 9.12.

Figure 9.12

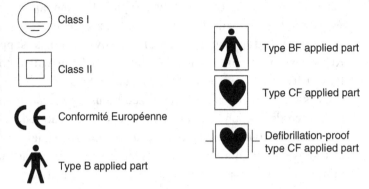

Examples of stickers attached to a medical device to identify its class and type. 'Defibrillation proof' means that the device can remain attached to the patient even if cardiac defibrillation is required.

9.4.1 Classes of Equipment

The **class of equipment** refers to the method by which the device is protected against potential electrical shocks to the patient.

Class I equipment is fitted with a three-core power cable that contains a protective earth wire. Any exposed metal parts on Class I equipment are connected to this earth wire. Class I equipment has fuses at the equipment end of the main power supply in both the live and neutral conductors. Should a fault develop inside the equipment and exposed metal come into contact with the main power supply, the earthing conductor conducts the fault current to ground. This current blows the fuse and breaks the circuit to remove the source of potential electric shock. According to safety standards the maximum resistance between the protective earth plug pin and conductive components is $0.2\ \Omega$.

Class II equipment is enclosed within a double-insulated case (or one layer of reinforced insulation) and does not require earthing conductors. Therefore Class II equipment is usually fitted with a two-pin plug. If an internal electrical fault develops in the device, it is unlikely to be hazardous as the double insulation prevents any external parts that might come into contact with the patient from having a high potential with respect to ground. Examples of Class II equipment include cardiac monitors.

Class III equipment is protected against electric shock by relying on the fact that no voltages higher than the safety extra low voltage (SELV), i.e. a voltage not exceeding 25 V AC or 60 V DC, are present within the instrument. This means that this type of equipment is either battery operated or supplied by a SELV transformer. Examples include implantable pacemakers and automated external defibrillators.

(a)

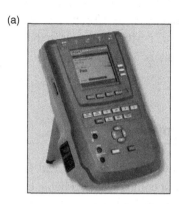

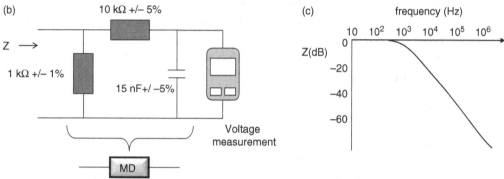

(a) A typical hand-held device used for electrical safety testing. (b) Internal circuit model (IEC60601) of the body. The abbreviation MD is used for medical device. (c) The frequency dependence of the impedance of the circuit model, which mimics the behaviour of the human body.

9.4.2 Types of Equipment

The **type of equipment** is defined in terms of the maximum permitted leakage current from the equipment. In Type B (body) equipment, which includes Classes I, II and III equipment, the maximum leakage current is 100 µA. This type of equipment cannot be directly connected to the heart. Type B applied parts may be connected to earth ground: examples of Type B equipment include medical lasers, MRI body scanners and phototherapy equipment.

Type BF (body floating) equipment refers to a floating, isolated component and is generally used for applied parts that have medium- or long-term conductive contact with the patient. Type BF equipment can be attached to the patient's skin but has a floating input circuit, i.e. there are no connections between the patient and the ground. Examples of Type BF equipment include blood pressure monitors.

Type CF (cardiac floating) equipment is the most stringent classification and is used for applied parts that may come into direct contact with the heart.

The maximum leakage current is 10 μA. The minimum resistance between the mains lead and ground is 20 MΩ, and between the mains lead and the components connected to the patient is 70 MΩ. Examples of type CF equipment include dialysis machines.

9.5 Safety Testing Equipment

It is estimated that up to 70% of all electrical faults can be detected by visual inspection, with the most common causes being cracks in the equipment housing, cuts in the electrical cables and fuses being replaced with ones with too low a rating. Nevertheless, regular maintenance of medical electrical equipment is needed to detect problems that cannot be diagnosed visually. A typical safety testing apparatus, an example of which is shown in Figure 9.13(a), can measure a number of parameters including:

  (i) line (mains) voltage

 (ii) ground wire (protective earth) resistance

(iii) insulation resistance

(iv) device current and lead (patient) leakage tests, including: ground wire (earth), chassis (enclosure), lead-to-ground (patient), lead-to-lead (patient auxiliary) and lead isolation (mains on applied part).

The leakage tests in (iv) are designed to simulate a human body coming in contact with different parts of the equipment. Since the apparatus cannot actually be connected to a human being, an electrical circuit is used to mimic the frequency-dependent impedance of the body, as shown in Figure 9.13(b). The circuit has a relatively constant impedance from DC to 1000 Hz, then a decrease of 20 dB/decade characteristic of a first-order RC circuit.

The line/mains voltage is in general either 110 volts or 240 volts, depending upon the convention in the particular country. Most medical equipment can operate with some variation above or below this number.

All Class I devices must be tested for the protective conductor resistance. The maximum resistances allowed are: devices with a removable mains power cable (0.2 Ω), devices including an in-built mains power cable (0.3 Ω), the mains power cable itself (0.1 Ω) and systems with multiple electrical outlets (0.5 Ω). Insulation resistance has no specific limits but the following values are used as guidelines. For Class I equipment, the insulation resistance should be greater than 2 MΩ (line to protective conductor), for Class II greater than either 7 MΩ (line to accessible conductive part of type BF application part), or 70 MΩ (line to type CF application part).

9.5.1 Leakage Current Measurements

Leakage currents are defined as those that are not part of the functional operation of the device. There are a number of different types of leakage current, which must be measured, and these are covered in the next section. In the description of leakage current measurements an applied part is defined as part of the medical equipment that comes into physical contact with the patient either by design or in practice. The limits for leakage currents are shown in Table 9.2.

Leakage currents are measured under both normal conditions (NC) and single-fault conditions (SFC). Normal conditions refer to cases in which all protections are intact and the device is operated in standby and full operational mode under normal use conditions. Normal conditions also include reversing the hot and neutral terminals on the mains supply, as this occurs commonly in daily practice, and applying a voltage to signal input/signal output (SIP/SOP) terminals such as USB or older RS-232 ports. There are a number of different SFCs, which are tested, including open-circuiting the protective ground, and open-circuiting each conductor on the mains supply, one at a time.

Connection for most tests is straightforward, with the MD being connected to the conductive point being tested. For medical equipment with a metal enclosure, the MD is connected to an unpainted portion of the enclosure. For measurements on a piece of equipment with an enclosure made of an insulating material, a piece of conductive metal foil of dimensions ~20 × 10 cm (simulating the size of the palm) is placed on the surface of the equipment and is connected to the MD. The foil is then tested in different positions on the insulating surface to determine the highest value of leakage current.

Several different types of leakage currents must be tested, as described in the following sections.

9.5.1.1 Earth and Enclosure Leakage Currents

The **earth leakage current** flows from the protective earth of the medical equipment through the patient back to the protective earth conductor of the power cord. This represents the total leakage current from all protectively earthed parts of the product. Measurement of this type of leakage current is required for Class I devices. Figure 9.14(a) illustrates the basic measurement of earth leakage current from a piece of medical equipment that uses a standard detachable power cord.

The **enclosure leakage current** represents the current that would flow if a person were to come into contact with the housing or any accessible part of the equipment.

Set ups for measuring: (a) earth leakage current, and (b) and (c) enclosure leakage currents.

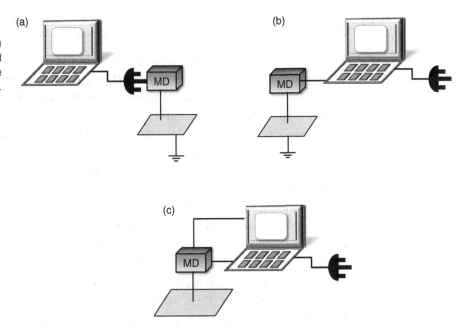

It is measured between any part of the enclosure and earth, Figure 9.14(b), and also between any two parts of the enclosure, Figure 9.14(c). Enclosure leakage currents are measured under NC and SFCs in which one supply conductor at a time is interrupted and, if present, the protective earth conductor is opened.

9.5.1.2 Patient Leakage Current: Normal and F-Type

The **patient leakage current** is very important to check since it directly affects patient safety. It originates from an unintended voltage appearing on an external source. The patient leakage current limits, shown in Table 9.2, are set so that the probability of ventricular fibrillation is less than 0.2%. Depending upon the type (B, BF or CF) of applied part, there are different requirements for how the leakage tests are performed and the type of fault conditions that must be included in the tests:

(i) the leakage current for Type B applied parts is measured between all similar applied parts tied together and ground, as illustrated in Figure 9.15(a)

(ii) Type BF applied parts are separated into groups that have different functions. The leakage current is measured between all applied parts with a similar function and ground as shown in Figure 9.15(b), and

(iii) the leakage current for Type CF applied parts must be measured from each applied part to ground individually, as shown in Figure 9.15(c).

Table 9.2 Maximum leakage currents (mA) for different medical devices under normal conditions (NC) and single fault conditions (SFC) (from IEC 60601–1)

	Type B		Type BF		Type CF	
	NC	SFC	NC	SFC	NC	SFC
Earth	0.5	1	0.5	1	0.5	1
Enclosure	0.1	0.5	0.1	0.5	0.1	0.5
Patient (DC)	0.01	0.05	0.01	0.05	0.01	0.05
Patient (AC)	0.1	0.5	0.1	0.5	0.01	0.05
Patient (F-type)	–	–	–	5	–	0.05
Patient (mains on SIP/SOP)	–	5	–	–	–	–
Patient auxiliary (DC)	0.01	0.05	0.01	0.05	0.01	0.05
Patient auxiliary (AC)	0.1	0.5	0.1	–	0.01	0.05

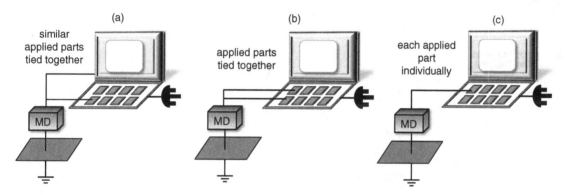

Figure 9.15 Schematic of the measurement protocols for patient leakage currents for: (a) Type B, (b) Type BF and (c) Type CF devices.

As in the previous test, SFCs correspond to interrupting one supply conductor at a time as well as open circuiting the protective earth conductor, if present.

Type F applied parts have an additional IEC 60601–1 requirement. The leakage current of each applied part is measured while applying 110% of the nominal mains/line voltage through a current-limiting resistor, as shown in Figure 9.16(a). During this test the signal input and the output parts are tied to ground. The leakage current is measured for normal and reversed mains polarity. Type B applied parts must have the additional SFC of 110% of the nominal mains voltage applied to all signal input parts (SIP) and signal output parts (SOP) during patient leakage measurement, as shown in Figure 9.16(b).

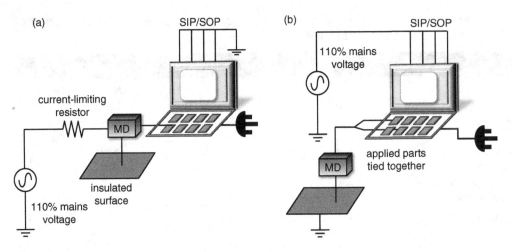

Figure 9.16 (a) Set up for additional required measurements for Type F applied parts. (b) Set up for additional required measurements for Type B applied parts.

Figure 9.17

Measurement set up for estimating the patient auxiliary leakage current for a piece of medical equipment.

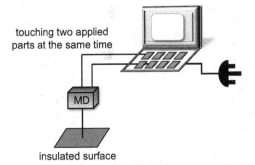

9.5.1.3 Patient Auxiliary Leakage Current

The **patient auxiliary leakage current** flows between any single applied part and all other applied parts tied together, as shown in Figure 9.17. Patient auxiliary leakage current is measured under NC as well as SFCs.

9.5.2 Earthbond Testing

Earthbond testing, which is also referred to as **groundbond testing**, is used to check the integrity of the low-resistance connection between the earth connector and any conductive metal parts that may become live if there is a fault on a piece of Class I medical equipment. The test involves applying a test current of 25 amps (or one-and-a-half times the maximum current rating of the device, whichever is higher) at 50/60 Hz between the earth pin of the mains supply plug and any metal part of the medical equipment that is accessible to the patient, including the earth reference point. As outlined earlier, faults in the detachable power cord account for the vast majority of earthbond testing failures. Resistance measurements are also performed, with the maximum value set at 0.1 Ω for fixed power cords and 0.2 Ω for equipment with detachable power cords.

9.6 Safety of Implanted Devices

Devices that are implanted into the body for extensive periods of time have a number of additional safety requirements compared to external devices. The most important one is a measure of its long-term biocompatibility, which involves the interactions between the materials forming the device (both the housing and the internal components) and the body. This is a two-way process: the device must not cause any toxicity to the body, and interactions with the body should not degrade the performance of the device. Three different areas are considered in this section: (i) the biocompatibility of different materials used in biomedical devices and the biological tests used to assess this; (ii) the electromagnetic safety of implanted devices that must receive and transmit signals through the body from/to an external device; and (iii) the safety of implanted devices in patients who must undergo diagnostic imaging studies and, in particular, magnetic resonance imaging.

9.6.1 Biocompatibility

The biocompatibility of a device or instrument is a very general concept, and there is no universally accepted definition, or indeed a standard required set of tests to demonstrate the biocompatibility of a given device for regulatory bodies such as the FDA. However, the regulatory body does need to be convinced that the device is safe to operate, and therefore the manufacturer has to perform as many targeted and well-characterized studies as possible, studies that best demonstrate the lack of adverse biological response from the materials used in the device, and the absence of adverse effects on patients. The extent of testing depends on numerous factors, which include the type of device, its intended use, the degree of liability, the potential of the device to cause harm and the intended duration for which the device will remain within the body. Some materials, such as those used in pacemakers, must remain viable and safe for many years or decades. In this case, a biocompatibility testing programme needs to show that the implant does not adversely affect the body during the long period of use.

In general, one can define biocompatibility as a measure of a chemical and physical interaction between the device and tissue, and the subsequent biological response to these interactions. Biological tests, covered in section 9.6.1.2, should be performed under conditions that mimic as closely as possible the actual use of the device, or material used in the device, and should be directly related to its specific intended use. It should be recognized that many classical toxicological tests developed, for example, to test the safety of pharmaceuticals, have to be adapted for testing the biocompatibility of materials (which are generally solids of many different geometries and sizes) and complete medical devices. Overall guidelines for biocompatibility testing can be found in International Standard ISO 10993–1, 'Biological Evaluation of Medical Devices – Part 1: Evaluation and Testing Within a Risk Management Process'. Figure 9.18 shows a general flowchart used by manufacturers to estimate what degree of biocompatibility testing is necessary before submission of a device for regulatory approval.

9.6.1.1 Materials Used in Implanted Devices

In general, there are three broad classes of materials that are used for biomedical devices: metals, ceramics and polymers [1]. Metals are used for the housing of cardiovascular devices, such as pacemakers, and also for the electrodes and leads. Ceramics are used in many types of sensors. Polymers are often used to coat metallic elements, such as pacemaker leads. Within the device itself, microfabricated components can contain of all three types of material.

Metals

Metals must be non-corrodable, sterilizable, high purity and (ideally) compatible with diagnostic scanning techniques, such as MRI, which means non-ferrous. Type 316 L stainless steel, containing approximately 18% chromium and 14% nickel, is by far the most common metal used in implant components where strength is needed; when the implant must remain in the body for a long time, molybdenum can be added to the stainless-steel alloy to form a protective layer to reduce corrosion due to an acidic environment. The other very commonly used metal is titanium and its alloys, most commonly Ti–6Al–7Nb (6% aluminium, 7% niobium) or Ti–6Al–4 V (6% aluminium, 4% vanadium), due to their very high tensile strength. Titanium alloys are used in the housing for many cardiovascular implants, such as pacemakers. Nitinol is another titanium alloy, containing roughly 50% titanium and 50% nickel, that is used in a wide range of applications. It is a 'superelastic' material (with greater than ten times the

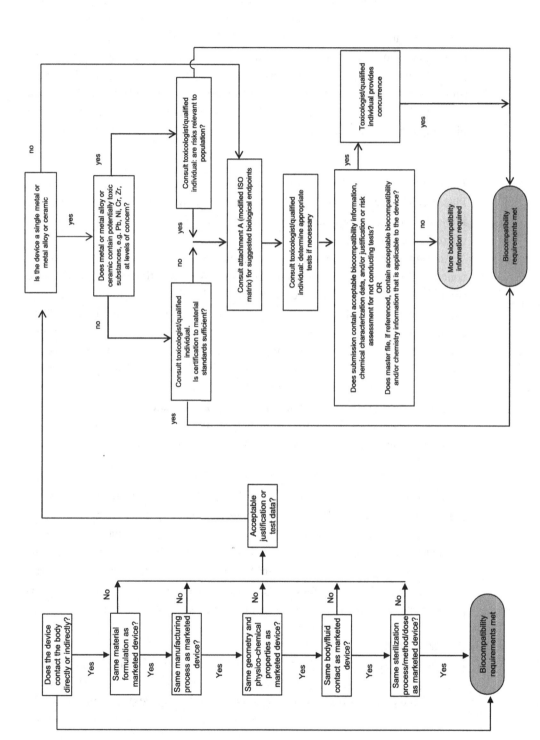

Figure 9.18 An example flowchart of the types of decisions that need to be made by a company in setting up a biocompatibility testing regime for a new device.

elasticity of stainless steel) that can return to its original shape after a significant deformation, which makes it ideal for medical devices that require large recoverable deformations, such as cardiovascular and gastro-intestinal stents. One use of nitinol was shown in section 8.4.1.4 describing an implanted pulmonary artery pressure sensor. Cobalt–chromium alloys (63 to 68% cobalt, 27 to 30% chromium, 5 to 7% molybdenum and trace nickel) are commonly used in artificial joints including knee and hip joints, due to their high wear resistance and biocompatibility, as well as in the manufacture of stents and other surgical implants.

Ceramics

Piezoelectric ceramics such as lead zirconate titanate are used in a range of implanted medical components, including piezoceramic sensors, actuators and micropumps.

Although ceramic components of an implanted device do not come into direct contact with tissue, it is important to assess their biocompatibility in case of device rupture. Several ceramics have been specifically developed for implants, one example being oxinium-oxidized zirconium, which has an oxidized zirconia ceramic surface, and is used in a ceramic knee implant. Compared to conventional cobalt–chromium alloys, it has approximately half the friction and lasts for a longer time. It can also be used in patients who have nickel allergies. Another example of biocompatible ceramics are coatings based on hydroxyapatite, a ceramic material that has a similar composition to natural bone, that are used in devices that release drugs after implantation of a coronary stent. Very thin films of this ceramic are coated onto a conventional wire mesh stent.

Polymers

Many different polymers, including polypropylene, polyethylene, polyamide, polytetrafluoroethylene (PTFE), polydimethylsiloxane, parylene, SU-8, silicone, polymethylmethacrylate and polyester, are used for cardiovascular, neurological, ophthalmic and subcutaneous devices. Silicones have very low surface tensions, extreme water repellency, high lubricating properties, are gas permeable and are highly resistant to degradation. The leads in cardiovascular devices, such as pacemakers, are coated with biocompatible polymers such as PTFE.

9.6.1.2 Biological Testing

In order to test biocompatibility, a number of biological tests are performed, including cytotoxicity, haemocompatibility, the effects of implantation, carcinogenicity and degradation, each of which is detailed in the next sections.

Cytotoxicity

The term cytotoxicity refers to cell death in mammalian cell cultures as a consequence of exposure to a harmful chemical. In-vitro cytotoxicity tests often use mammalian fibroblasts, which can be seeded onto the surface of different components of the device. After incubation in an appropriate medium, the medium is removed and the device components washed. The number of live and dead cells can be assessed using various commercial viability/cytotoxicity assay kits. In-vivo cytotoxicity tests in animals use a variety of techniques including haematoxylin and eosin (H&E) staining of histological slices to determine the degree of neovascularization and the presence of inflammatory cells. Immunofluorescence staining can be used for the detection of macrophages. Although cytotoxicity testing is a very useful screening tool, there are several caveats to interpreting the results. The first is that the correlation of results from these assays with observed effects in humans is highly variable [2]. In addition, the degree to which a particular organ can function for a given degree of cytotoxicity is also highly variable: for example, epithelial cells can rapidly regenerate whereas neuronal tissue does not have this capability.

Haemocompatibility

Haemocompatibility is a measure of the interactions of a device with blood. These interactions range from transient haemolysis and minimal protein adsorption to much more serious coagulation and cell destruction. The FDA and ISO requirements for devices not implanted into the vascular system can be met using an in-vitro haemolysis test. However, devices that are intended to be implanted in the vascular system must be evaluated in vivo, with recommendations for five test parameters: thrombosis, coagulation, platelets, haematology and immunology. Specific measures for each of these tests include the following:

(i) Thrombosis: per cent occlusion; flow reduction; autopsy of device (gross and microscopic), autopsy of distal organs (gross and microscopic).
(ii) Coagulation: partial thromboplastin time (non-activated), prothrombin time, thrombin clotting time; plasma fibrinogen, fibrin degradation product.
(iii) Platelets: platelet count, platelet coagulation.
(iv) Haemotology: leukocyte count and differential; haemolysis (plasma haemoglobin).
(v) Immunology: in-vitro measurement of the levels of C3a, C5a, TCC, Bb, iC3b, C4d and SC5b-9.

Implantation effects

The Unites States Pharmacopeia 2007 test is the reference method for both the FDA and ISO (10993–6) for testing the macroscopic effects of device implantation. The test can be performed for either short-term (from 1 up to 12 weeks) or long-term (from 12 to 104 weeks) implantations. The test is designed to evaluate the effects of plastic and other polymeric materials that come into direct contact with living tissue. Narrow strips of plastic are implanted using a hypodermic needle into the paravertebral muscles of rabbits under anaesthesia. The test is considered successful (i.e. the material is biocompatible) if, in each rabbit, the reaction to not more than one of the four samples is significantly greater than that to the strips of USP Negative Control Plastic RS. Macroscopic visual checks of the tissue around the implants are performed to detect haemorrhage, necrosis, discolouration, infection or encapsulation. Additional histopathology tests can be performed to determine the effects (if any) on connective tissue proliferation, mononuclear cell infiltration, polymorphonuclear cell infiltration, muscle degeneration and multinucleated giant cell infiltration.

Carcinogenicity

Carcinogenicity studies are not very often required for medical devices. The ISO standard states that 'carcinogenicity tests should be conducted only if there are suggestive data from other sources'. Only one animal model needs to be tested, most commonly Sprague–Dawley rats, for two years. A device is implanted into an animal by making a surgical incision into the flank, opening a pouch, inserting the device and suturing the pouch closed. In control animals, only the surgical procedure is performed with no device inserted. The main parameters assessed are survival time and the occurrence of any tumours. Other measures include various urine parameters (colour, pH, ketone and urobilinogen levels, the specific gravity/ refractive index, albumin, glucose, occult blood, urinary sediment and bilirubin) collected prior to the test starting, at six-month intervals during the study and just prior to the final sacrifice. Clinical pathology and haematology measurements are made on blood samples collected at the same intervals.

Effects of material degradation

All materials degrade to some degree within the body, and the particular processes vary for different materials. In metals, corrosion can result in the release of metal ions or other compounds that lead both to local tissue damage and also to the device performance being reduced. Reactive metals such as titanium and chromium react with oxygen to form a metal oxide layer, which acts as a barrier to diffusion of metal ions and prevents corrosion. Damage to this layer in vivo may greatly accelerate the degradation process. Surface interactions with proteins or bacteria that alter local pH or reduce the availability of oxygen may compromise the stability of an oxide layer. Despite the high strength and hardness of metal alloys commonly used in implantable devices, some wear does occur, which can produce cytotoxicity [3]. Numerous investigators have demonstrated significant decreases in cell viability upon direct contact with metal particles in vitro. Metallic implant materials are most commonly evaluated in phosphate-buffered saline, Hank's solution or Ringer's solution, which contain physiologically relevant salts. Polymers that come into contact with tissue or body fluids can undergo chemical degradation via hydrolysis (reaction with water producing chain cleavage), oxidation and enzymatic processes. Functional groups that are highly susceptible to hydrolysis include esters, amides, urethanes and carbonates [4].

As an example of a report submitted to the FDA for a new continuous glucose monitoring device (covered in Chapter 8), the following section describes the essential points of the biocompatibility testing provided to the authorities in a successful application.

Biocompatibility Testing

Testing was performed for the sterilized sensor that is inserted into the skin. Other components that come into contact with the user's skin surface are made from materials with established biocompatibility, including the introducer needle and the sensor mount adhesive patch.

Sensor biocompatibility was performed following ANSI/AAMI/ISO 10993–1 Standard (1997 Edition). All test protocols were performed following Good Laboratory Practice (GLP) as described in 21CFR Part 58. The results of the biocompatibility testing for the system sensor component are listed below:

Cytotoxity (MEM elution)	Pass – non-cytotoxic
Sensitization (maximization test)	Pass – no evidence of sensitization
Irritation (ISO intracutaneous)	Pass – primary irritation responses negligible
System toxicity (USP/ISO systemic injection test)	Pass – non-toxic
Subchronic implantation irritancy	Pass – no gross evidence of local irritancy
Genotoxicity (Ames test)	Pass – non-mutagenic
Haemocompatibility (haemolysis in vitro)	Pass – non-haemolytic

Sterility

The electron-beam sterilization process used to sterilize the sensor assembly was validated according to the requirements of ISO 11137 and AMI TIR27 Method VDmax.

Sensor Shelf Life

Sensor shelf life has been determined through real time (25°C) and accelerated studies. Under dry storage conditions, sensor performance has been shown to be acceptable for a shelf life of six months when stored at 2 to 25°C. Additional studies were carried out to verify that the packaging materials maintain a suitable dry environment for at least six months and have no deleterious effect on sensor stability.

Packaging and Shipping Testing

Final systems and SDU kits were packaged using standard materials and methods, and subjected to shipping tests as per ASTM D 4169–05 including altitude at 12 000 feet. All samples passed functional testing performed at the conclusion of the test.

9.6.2 Electromagnetic Safety

In addition to determining that an implanted device is safe from a biological point of view, it is also critical to ensure that an active device is safe when operating in normal mode. For wireless implanted devices this means that the amount of power transferred from the external source to the device, and vice versa through the body, is safe. As discussed in section 8.3.1.2 the power absorbed by the body is proportional to the square of the electric field produced by the transmitting antenna and directly proportional to the tissue conductivity. The IEC and ICNIRP limit the SAR to 2 W/kg over 10 g of contiguous tissue. If an external antenna is used to transfer power to the implanted device, then this must also be taken into consideration. Testing is performed both with electromagnetic simulations and also via thermal measurements using materials that have the same conductivity and dielectric properties as tissue.

For the FDA application for a continuous glucose monitor, the following information was supplied in terms of electromagnetic safety.

EMI/EMC/ESD and Product Safety Testing

A sequence of EMI/EMC/ESD and product safety testing for the system receiver and transmitter was performed by external accredited laboratories. The ESD/RF immunity testing showed that the transmitter and receiver units functioned correctly after exposure to electromagnetic fields. Emissions from the transmitter and receiver were also tested for compliance and passed. Testing was conducted according to the following standards:

a. CISPR 11 (requirements for medical devices)
b. IEC 60601–1-2
c. FCC Part 15 subpart B
d. EN 300 220–3
e. EN 301 489–3

Additional immunity testing was conducted beyond the standards listed to cover specific frequencies as follows: frequency in MHz, 124, 161.5, 315, 836.5, 902.5, 915, 1747.5, 1880 and 2450.

Additional special immunity testing was conducted beyond the listed standards to cover interference generated by electronic article surveillance devices and metal detectors.

Electrical safety was tested for compliance to the standards:

a. UL 60601–1
b. IEC 60601–1-1

FCC regulatory testing for the system transmitter was conducted, according to the requirements prescribed by the design specification as they relate to electromagnetic emissions and compliance with the applicable FCC standards as listed in CFR 47, Part 15.231. The testing for the system transmitter was performed by an external accredited laboratory and the transmitter complies with the standards.

9.6.3   Clinical Studies

If a regulatory body such as the FDA deems that a clinical trial is necessary to approve a new Class III device, then a company must present a specific hypothesis to be tested (or question to be answered) and design a study that will answer that specific hypothesis or question. Having come up with a hypothesis, the following questions need to be considered in designing the clinical trial.

(i) With what treatment (control) should results from the investigational device be compared?

(ii) Is the objective of the study to show that the new device performs better than or equivalently to an alternative therapy?

(iii) How many patients (sample size) are required to detect a clinically meaningful difference between the treatment and control groups with adequate statistical power, or ensure an acceptable degree of similarity?

(iv) What is the diagnostic criterion by which to select study subjects?

(v) How will patients be assigned to the treatment and control groups?

(vi) What baseline characteristics (characteristics of patients at the time of entry into the study) are important to prognosis?

(vii) What will be the primary end point(s) of the study?

(viii) What are the individual patient success/failure criteria?

(ix) Will the patient, physician or another person be responsible for treatment administered?

(x) What follow-up information is important to collect?

(xi) How will the safety of the device be established?

9.7 Design of Devices That Can Be Used in a Magnetic Resonance Imaging Scanner

Many patients with implanted devices must also undergo a number of diagnostic tests to assess treatment progress. Issues of compatibility of an implanted medical device with these tests may well arise. One of the most challenging environments for an implanted medical device is that in an MRI scanner. In the past many implanted devices such as pacemakers and cochlear implants were considered to be MRI incompatible, i.e. the patient could no longer undergo an MRI scan unless the implant were removed. Much effort has been made over the past few years to make such devices 'MRI-conditional', i.e. an MRI can be performed under well-defined scanning conditions.

As an example of such engineering developments in device design, consider the pacemaker and related devices that have been extensively described in Chapter 8. New devices have recently been specifically designed by the major companies (Medtronic, St. Jude Medical, Biotronik) that can be scanned using MRI. Potential effects of the MRI system on pacemakers and ICDs are summarized in a number of papers [5, 6] and can be separated into three broad classes.

(i) The very strong static magnetic field (B_0) of an MRI scanner (typically 1.5 or 3 tesla) produces a magnetic force and torque on any ferromagnetic materials. Many conventional implantable pacemakers contain a static magnetic field sensor, a reed switch, MEMS sensor, or giant magnetoresistance sensor, which is used to inactivate the sensing function of a pacemaker. A clinical MRI has a B_0 field strong enough to activate the implantable pacemaker's magnetic sensor, causing the pacemaker to revert to asynchronous pacing, which can lead to tachycardia and ventricular fibrillation should the pacemaker fire in the 'vulnerable phase' of the cardiac cycle. A pacemaker can also create image artefacts in the MR scan due to magnetically induced field distortions from the pulse generator. The primary factors that affect the degree of image artefact are the magnetic susceptibility and the mass of the materials used in the pulse generator.

Solutions: The reed switch can be replaced by a Hall sensor. The pacemaker housing can be made from a conductive material, such as titanium, 316 L stainless steel or other similar materials, which do not contain any ferrous components.

(ii) The effects of the pulsed gradient field in the MRI scanner are to induce currents in the loop area (on the order of 200 to 300 cm^2) defined by the pacemaker lead and return path from the distal pacing electrode back to the implanted subcutaneous pulse generator. Induced currents and voltages in a pacemaker can cause inappropriate sensing and triggering, and even stimulation. The induced voltage can be in the order of 500 mV peak to peak.

Solutions: the size of the pacemaker loop can be made much smaller, and a band-stop filter can be placed in the pacemaker lead to reduce the pick up from the gradients (which is in the kHz frequency range) while allowing the DC pacing voltage to pass through.

(iii) The RF field can result in tissue heating at the site of the electrode tip. A clinical MRI can transmit a peak power of 30 to 60 kW at 64 MHz for a 1.5 T MRI system or 128 MHz for a 3 T system. Depending upon the length of the pacemaker lead (see Problems) large increases in temperature can be induced at the end of the pacemaker lead.

Solutions: RF filters are added to the pacing lead tip to prevent RF currents from heating and damaging tissue: the filters do not affect either sensing signals to the pacemaker or pacing signals from the pacemaker, which are transmitted through the lead. The leads can also be shielded. Modified filter boards are added to the device to reduce the risk of cardiac stimulation due to rectification.

In clinical practice, an external 'MRI-activator device' is used to put the pacemaker in its 'MRI settings' state: after the scan has finished the activator is used to disable the MRI settings and restore normal pacemaker operation. Tests have compared the safety and efficacy of the MRI-conditional pacemakers with a conventional DDD pacemaker. Because the redesign of the MRI-conditional system required increased lead diameter and stiffness, implantation success was also evaluated. No clinically relevant differences in lead performance were found between the two types of pacemaker [7, 8]. Randomized trials have also shown that there were no complications reported among subjects who underwent scans, and there were no MRI-attributed sustained ventricular arrhythmias, asystole or pacemaker malfunctions [9].

Appendix: Safety Policy Documents

1. IEC 60601–1, Medical Electrical Equipment – Part 1: General Requirements for Safety and Essential Performance (Geneva: International Electrotechnical Commission, 1995)
2. IEC 60601–2-31 Ed. 2.0 b:2008 Medical Electrical Equipment – Part 2–31: Particular Requirements for the Basic Safety and Essential Performance of External Cardiac Pacemakers with Internal Power Source
3. National Fire Protection Association, NFPA 99, Standard for Health Care Facilities (Quincy, MA: NFPA, 2002)
4. ANSI/AAMI ES1:1993, Safe Current Limits for Electromedical Apparatus (Arlington, VA: AAMI, 1993)
5. BS 5724: Electrical Safety of Medical Equipment

PROBLEMS

9.1 Explain the physiological basis for the sensation threshold shown in Figure 9.4. What happens just above this threshold?

9.2 Similarly, explain what happens physiologically at the threshold of the let-go limit current.

9.3 Explain the general frequency characteristics of the let-go limit and sensation threshold currents, i.e. why they are high at very low frequencies and also very high frequencies.

9.4 The figure on the left shows the set up for electrocautery, a procedure in which tissue is heated by an electric current. What are the dangers in the figure on the right where there is poor contact between the patient and the return path?

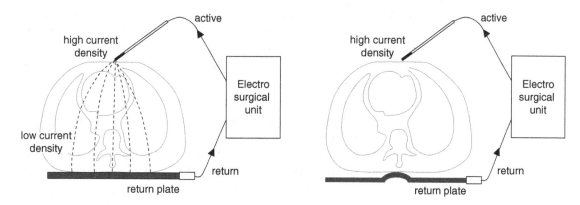

9.5   Identify five medical devices that are in Classes I, II and III. Explain briefly why each device fits into the particular category.

9.6   Suggest a situation corresponding to each of the following current measurements:
 (i) patient auxiliary leakage current (Figure 9.17)
 (ii) patient leakage current for Type B, Type BF and Type CF (Figure 9.15)

9.7   A bed lamp is at a potential of 240 volts, and the insulation has become frayed so that this potential appears at the outside. The patient comes into contact with the lamp. A saline-filled catheter with a resistance of 100 kΩ is connected to the patient's heart. There is a 10 MΩ leakage resistance and 100 pF capacitance in the insulation between the ground and the saline in the sensor. If the skin impedance is high, 1 MΩ, is there a hazard for microshock? How about if the patient's hands are wet, and the impedance drops to 100 kΩ?

9.8   The purpose of a cardiac defibrillator is to reset the cardiac pacemaker cells by inducing a macroshock into the system. Design a system that can do this. How much current should be provided, and for how long? How should it be connected? What are the risks associated with the design?

9.9   Related to question 9.8, some equipment is designed to be defibrillation proof. How might this be performed? Determine what the power handling requirements of such a circuit might be.

9.10   An RDC and GFCI circuit basically differ only in the value of the trip current needed. Investigate why these devices have different standards?

9.11   Explain how the circuit for a GFCI shown in Figure 9.11 works.

9.12   In section 9.6.2 the following additional tests were performed: 'Additional immunity testing was conducted beyond the standards listed to cover specific frequencies as follows: frequency in MHz, 124, 161.5, 315, 836.5, 902.5, 915, 1747.5, 1880 and 2450. Additional special immunity testing was conducted beyond the listed standards to cover interference generated by electronic article surveillance devices and metal detectors.' Determine why these particular frequencies were chosen.

9.13   Considering the relatively new leadless pacemaker considered in section 8.4.1.6, suggest which biological, clinical and electromagnetic safety tests would have had to be performed for approval.

9.14 One of the major issues with using medical devices in an MRI is potential heating of either the device or the wires that are attached to the device. For a pacemaker, what length of pacemaker lead gives the highest degree of potential heating at the tip on a 3 tesla clinical magnet? What is the effect of the pacemaker lead being surrounded by tissue of this length?

9.15 Following from question 9.14, is the danger of heating greater for a completely bare wire, an insulated wire or a wire such as a pacemaker lead, which is insulated but has a bare tip? Explain your reasoning.

9.16 Although both aluminium and titanium are non-magnetic, there is the chance that such implants can experience large forces when a patient moves into an MRI scanner. Explain what gives rise to these forces.

9.17 One of the medical devices that is most problematic with regards to MRI is a cochlear implant, which often has to be removed for the patient to be scanned. By searching the literature, discuss briefly some of the new advances that have been proposed to devise an MRI-conditional device. What are the trade-offs, if any, with the new design?

REFERENCES

[1] Khan, W., Muntimadagu, E., Jaffe, M. & Domb, A. J. *Implantable Medical Devices*. Controlled Release Society; 2014.

[2] Wilsnack, R. E. Quantitative cell culture biocompatibility testing of medical devices and correlation to animal tests. *Biomater Med Devices Artif Organs* 1976; **4**(3–4):235–61.

[3] Savarino, L., Granchi, D., Ciapetti, G. *et al.* Effects of metal ions on white blood cells of patients with failed total joint arthroplasties. *J Biomed Mater Res* 1999; **47** (4):543–50.

[4] Ratner, B. D. *Biomaterials Science: An Introduction to Materials in Medicine*. Academic Press; 2004.

[5] Colletti, P. M., Shinbane, J. S. & Shellock, F. G. 'MR-conditional' pacemakers: the radiologist's role in multidisciplinary management. *Am J Roentgenol* 2011; **197**(3):W457–9.

[6] Shinbane, J. S., Colletti, P. M. & Shellock, F. G. Magnetic resonance imaging in patients with cardiac pacemakers: era of 'MR Conditional' designs. *J Cardiovasc Magn Reson* 2011; **13**:63.

[7] Forleo, G. B., Santini, L., Della Rocca, D. G. *et al.* Safety and efficacy of a new magnetic resonance imaging-compatible pacing system: early results of a prospective comparison with conventional dual-chamber implant outcomes. *Heart Rhythm* 2010; **7**(6):750–4.

[8] Wollmann, C. G., Thudt, K., Vock, P., Globits, S. & Mayr, H. Clinical routine implantation of a dual chamber pacemaker system designed for safe use with MRI:

a single center, retrospective study on lead performance of Medtronic lead 5086MRI in comparison to Medtronic leads 4592–53 and 4092–58. *Herzschrittmacherther Elektrophysiol* 2011; **22**(4):233–6; 239–42.

[9] Wilkoff, B. L., Bello, D., Taborsky, M. *et al.* Magnetic resonance imaging in patients with a pacemaker system designed for the magnetic resonance environment. *Heart Rhythm* 2011; **8**(1):65–73.

Glossary

Accelerometer: a transducer which measures acceleration, typically in all three directions, and produces a proportional output signal. Accelerometers are used, for example in remote patient monitors to measure the amount of physical activity of a patient rehabilitating at home.

Acoustic feedback: a phenomenon most commonly encountered as a whistling sound produced by a hearing aid, which typically occurs at frequencies between 1.5 and 8 kHz. Feedback arises from a component of the sound, produced by the receiver, which escapes from the ear canal and is subsequently picked up by the microphone and re-amplified.

Action potential: the change in electrical potential between the inside and outside of a cell that occurs when it is stimulated.

Alpha waves: electroencephalography signals with frequencies between 8 and 12 Hz which are typically recorded by electrodes placed over the occipital region in the brain.

Amplitude shift keying (ASK): a digital modulation technique for wireless data transmission from a device which may be implanted in the patient. The amplitude (either one or zero) of the transmitted signal is modulated by the digital output of the analogue-to-digital converter.

Analogue-to-digital converter (ADC): an electronic circuit used to digitize an analogue signal, the output being a series of ones and zeros. Different architectures are used depending upon the required resolution and speed of the ADC.

Armature: a structure which produces an electric current when it rotates in a magnetic field. Armatures are used extensively in receiver design for digital hearing aids.

Atrial fibrillation: a cardiac condition manifested by rapid and irregular beating of the heart which can be caused, for example by heart failure, various diseases of the coronary artery or heart valves, and hypertension.

Attack time: the attack time in a digital hearing aid compressor circuit is defined as the time taken for the output to stabilize within 2 dB of its final value after the input sound pressure level increases from 55 to 80 dB.

Automatic gain control (AGC): a term used in digital hearing aids referring to the gain of the device being set automatically for different listening situations, i.e. when soft or loud sounds occur. AGCs are also called fast-acting compressors or syllabic compressors since the changes in gain occur on a timescale comparable with the durations of individual syllables in speech.

Beta waves: electroencephalography signals with frequencies between 12 and 30 Hz which can be recorded from electrodes placed over the frontal and central regions of the brain.

Biopotential: bipotentials within the body correspond to ionic voltages produced as a result of the electrochemical activity of excitable cells. They are measured using transducers (usually surface electrodes) to convert ionic potentials into electrical potentials. Measurements of biopotentials include electrocardiography (heart), electromyography (muscle), electrooculography (eye) and electroencephalography (brain).

Bispectral index (BIS): an electroencephalographic-based measurement on a patient which gives an indication of the depth of anaesthesia. The index runs from 0–100 (zero corresponding to no electrical activity) with a value between 40 and 60 being appropriate for general anaesthesia. The measure is an empirical one, based on a weighted combination of time-domain, frequency-domain and high order spectral subparameters. The BIS is currently the only quantitative measure that is approved by the Food and Drug Administration for monitoring the effects of anaesthesia on brain activity.

Bode plot: a very simple and quick method for plotting the approximate frequency response of an electrical circuit in terms of its magnitude and phase. It was originally devised by the Dutch engineer Hendrik Wade Bode in the 1930s.

Body protected areas: areas in a hospital designed for medical procedures in which patients are connected to a piece of equipment or a device, such as an intravenous catheter, which lowers the natural resistance of the skin.

Bradycardia: a slower-than-normal heart rate which can be caused by many conditions including heart tissue damage, congenital heart defects, various inflammatory diseases, or the response to high blood pressure medicines.

Brain–computer interfaces: systems which connect the electrical activity of the brain to a computer-controlled external device. An example is an interface which allows a severely physically-challenged patient to control an external device based upon signals from implanted electrodes.

Buffer amplifier: an operational amplifier configuration with unity gain which buffers a circuit placed at its output from changes in impedance of a circuit connected to its input.

Burst suppression: successive periods of normal or higher-than-normal amplitude EEG signals alternating with periods of very low signals close to the noise floor. Burst suppression can be caused by various etiologies of encephalopathy, and is also associated with general anaesthesia.

Butterworth filter: a form of filter (implemented either in hardware or software) which has a maximally-flat response in its passband. It was first described in 1930 by the British engineer Stephen Butterworth.

Cardiac protected areas: areas in a hospital in which the operating procedure involves placing an electrical conductor within or near the heart. Electrical protection against fibrillation arising from small leakage currents is required in these areas.

Cardioverter-debrillator: a battery-powered device placed under the skin of a patient that records heart rate. Thin electrodes connect the cardioverter-defibrillator to the heart and, if an abnormal heart rhythm is detected, they deliver an electric shock to the heart to restore normal operation.

Cellular depolarization: a process by which the electric charge distribution inside and outside a cell changes, resulting in a lower negative charge inside the cell. Cellular depolarization is the first step in an action potential being produced.

Chebyshev filter (Type I): a form of filter (implemented either in hardware or software) which has an equiripple response in the passband, and minimizes the difference between the ideal "top-hat" frequency characteristics and the actual response over the entire range of the filter. It was first described in 1931 by the German engineer Wilhelm Cauer, who used the theory of Chebyshev coefficients to design the filter.

Common-mode signals: signals which are present on all conductors either within a device or connecting different components of the device together. For example, common-mode interference signals can be induced in the ECG wires connecting electrodes to the amplifier unit via electromagnetic coupling with the electrical power system. Common-mode signals can be reduced by using a differential amplifier, whose output is the difference between the two input signals.

Comparator: an electrical circuit which compares the magnitude of an input voltage with a reference voltage and produces a binary output signal. If the input signal is greater than the reference then the output of the comparator is one value, if it is less than the reference voltage then the output is the other value.

Compression ratio: a characteristic of compression circuits in digital hearing aids, defined as the change in input sound pressure level required to produce a 1 dB change in output level.

Conformité Européene (CE) mark: a mark (displayed on all relevant medical devices) that indicates conformity with the legal and technical directives of the European Union.

Cross-correlation: a class of data processing techniques that are used to measure the "similarity" of two signals. These signals are often shifted in time with respect to one another, or have the same time-dependent behaviour but occur in different parts of the body.

Current-of-injury mechanism: a phenomenon referring to the current generated when an injured part of an excitable tissue (e.g. nerve or muscle) is electrically connected to an uninjured region. The current arises since the injured tissue is negatively charged with respect to the uninjured tissue.

Cytotoxicity: refers to cell death as a consequence of exposure to a harmful chemical. In vitro cytotoxicity tests of medical devices are often performed using mammalian fibroblasts, whereas in vivo cytotoxicity tests are performed in animals.

Decimation: a signal processing process in which M consecutive data points are averaged together to give a single point. Decimation is the final step in an over-sampling analogue-to-digital receiver. Since the quantization noise in consecutive data points is random, the overall quantization noise is reduced by a factor of $\sqrt{M}$ compared to that of a signal sampled at the Nyquist frequency.

Delta sigma ADC: an analogue-to-digital architecture which can be used for applications requiring high resolution (>16 bits) and sampling rates up to a few hundred kHz. Data is oversampled and then undergoes digital filtering and decimation.

Delta waves: electroencephalographic signals with frequencies below 4 Hz which have their sources either in the thalamus or in the cortex. A large amount of delta activity in awake adults is considered abnormal and is often associated with neurological diseases.

Deterministic signals: those for which "future" signals can be predicted based upon "past" signals. For example, electrocardiography signals are considered deterministic over a short time-frame, since neither the heart rate nor the shape of the electrocardiography trace changes significantly.

Digital filter: a software-implemented filter used to process data after it has been digitized. Digital filters can either be implemented in real-time or applied after data acquisition has finished.

Directivity factor/index: a measure of the directionality of a microphone or microphone array used in a digital hearing aid.

Driven right leg circuit: an electronic circuit incorporating negative feedback which is used to reduce the common-mode signal in an electrocardiography measurement.

Dynamic range: a characteristic of signals or measurement devices which represents the ratio of the maximum-to-minimum values of the signals produced, or the signal that can be digitized, respectively. Dynamic range is normally reported on a decibel scale.

Electrocardiogram (ECG): a measurement which uses up to ten electrodes placed on the surface of the body to measure the electrical activity of the heart. Up to twelve different voltages traces can be displayed on the monitor simultaneously.

Electroencephalogram (EEG): a surface-based measurement of the electrical activity of the brain which uses typically between 32 and 128 electrodes placed on the scalp of the patient.

Einthoven's triangle: a virtual triangle with vertices at the two shoulders and the pubis, which forms three limb leads (leads I, II and III) in electrocardiography. The heart is at the centre of this approximately equilateral triangle.

Electret: a key component of microphones used in digital hearing aids, consisting of a thin (about 12 μm) layer of dielectric which is glued on top of a metal back electrode and charged to a surface potential of approximately 200 volts.

Electronic health: a very broad term used to describe healthcare practices supported by electronic devices and communication.

Energy harvesting: a process which uses intrinsic energy sources within the body to power implanted biomedical devices. Examples of energy harvesting include patient movement and body heat.

Flicker noise: one of the basic sources of noise in electronic circuits (the others include Nyquist/Johnson and shot noise). Flicker noise occurs due to variable trapping and release of carriers in a conductor and can be modelled as a noise current.

Food and Drug Administration: an agency within the United States of America Department of Health and Human Services, responsible for "protecting the public health by assuring the safety, effectiveness, quality, and security of human and veterinary drugs, vaccines and other biological products, and medical devices".

Fourier transform: a mathematical operation which transforms, for example, time-domain data into the frequency domain. It can be used to characterize the frequency content of biosignals, and to design appropriate filters to remove unwanted interfering signals.

Frequency shift keying (FSK): a digital modulation technique used for wireless transmission of data from a biomedical device. In binary FSK modulation, two different frequencies are transmitted depending upon whether the input to the modulator is a zero or a one.

Functional electrical stimulation: a setup that uses low energy electrical pulses to generate body movements in patients who have been paralyzed due to injury to the central nervous system.

Goldman-Hodgkin-Katz (GHK) equation: relates the resting potential across a cell membrane to the intracellular and extracellular concentrations, as well as cellular permeabilities, of sodium, potassium and chloride ions.

Ground fault circuit interrupter: a device that shuts off an electric circuit when it detects current flow along an unintended path, such as through a patient to ground.

Helmholtz resonance: an acoustic phenomenon which occurs in the middle ear cavity, corresponding to the resonance of the air volume within the cavity (in combination with the earmold and vent). A typical resonance frequency is around 500 Hz.

Helmholtz double layer: refers to one layer of ions tightly bound to the surface of an electrode and an adjacent layer of oppositely-charged ions in solution, the effect of which is to cause the electrolyte to acquire a net positive, and the electrode a net negative, charge. The nature of the double layer affects the reaction kinetics at the electrode surface.

Instrumentation amplifier: an arrangement of three operational amplifiers which provides a very high common mode rejection ratio for differential inputs such as electrocardiographic leads.

Intraocular pressure (IOP): the fluid pressure inside the eye. Elevated ocular pressure is the most important risk factor for glaucoma. New micromachined sensors have been designed to measure IOP in situ.

Intrathoracic impedance: the impedance measured between the right ventricular lead and the housing of an implantable cardioverter-defibrillator. The impedance reflects the amount of fluid accumulation around the heart and lungs, and its value has been shown to decrease approximately 15 days before the onset of heart failure.

Investigational device exemption (IDE): the process by which preliminary clinical trials can be performed on devices that are only officially approved after a successful full clinical trial. For low-risk devices, the local institutional review board or medical ethics committee at the hospital or laboratory where testing is to be performed, can give such approval.

Knowles Electronics Manikin for Acoustic Research (KEMAR): the most commonly-used model for assessing the performance of hearing aids. The dimensions are based on the average male and female head and torso. The manikin design reproduces effects such as diffraction and reflection caused by the ears as sound waves pass by the human head and torso. A microphone is placed inside the manikin to measure the frequency-dependent response of the particular hearing aid being tested.

Lead voltage: the potential differences either between electrode pairs (bipolar) or between a single electrode and a virtual ground (unipolar) in an electrocardiography measurement.

Leakage currents: a current from a medical device which can potentially pass through the patient. There are many different categories of leakage current including earth/enclosure leakage currents, patient leakage currents and patient auxiliary leakage currents.

Macroshock: refers to the effects of currents on the order of hundreds of mA passing through the patient. Macroshocks can be caused by a patient touching a device at a very high potential.

Micro-electro-mechanical systems (MEMS): devices which consist of components between approximately 1 and 100 micrometres (i.e. 0.001 to 0.1 mm) in size, with the assembled MEMS devices generally ranging from 20 micrometres to 1 millimetre.

Microshock: the effect of a small current, as low as hundreds of micro-amps, which flows directly through the body via a highly conducting path such as a pacemaker wire or a central venous pressure catheter.

Mobile health: a general term for the use of mobile phones and other wireless technology in medical care.

Moving average filter: a data processing algorithm used to determine a general trend in the signal level. It is particularly useful when individual measurements are very noisy.

Myocardial ischaemia: occurs when blood flow, and therefore oxygen delivery, to the heart muscle is obstructed by a partial or complete blockage of a coronary artery.

Noise factor: the ratio of the signal-to-noise ratio at the input of an amplifier divided by the signal-to-noise ratio at its output.

Noise figure: the logarithm of the noise factor of an amplifier, measured in dB.

Nyquist/Johnson noise: noise in conducting materials which arises from temperature-induced fluctuations in carrier densities. The noise is proportional to the square root of the conductor resistance and the measurement bandwidth.

Nyquist-Shannon sampling rate: the Nyquist-Shannon sampling theory states that in order to correctly sample a signal that consists of frequency components up to a maximum of f_{max}, the sampling rate must be at least $2f_{max}$, otherwise signals alias.

Operational amplifiers: the most commonly-used electronic component to amplify small signals in biomedical devices. Op-amps are active devices which consist of networks of many transistors.

Oscillometric method: a method of measuring blood pressure non-invasively, which works on the principle that when the artery opens during the systolic/diastolic cardiac cycle, a small oscillation is superimposed on the pressure inside the cuff. The oscillation is detected using a piezoelectric pressure sensor.

Oversampling: a method which can increase the effective resolution of an analogue-to-digital converter. Oversampling involves sampling the signal at a much higher rate than required by the Nyquist-Shannon theorem. After digitization, the data undergoes digital filtering and is then decimated.

P-wave: the feature in the ECG which is produced by atrial depolarization of the heart.

Pacemaker cells: specialized cardiac cells in the sinoatrial node that are self-excitatory, i.e. they do not need an external trigger. Pacemaker cells produce an action potential which spreads through both atria.

Parkes error grid: a graph for assessing the performance of a particular blood glucose measuring system, based on the responses of 100 scientists who were asked to assign risk to self-monitoring blood-glucose systems for Type 1 and Type 2 diabetes patients. Values from the device are compared with a "gold-standard" reference value from a validated laboratory measurement.

Phase shift keying (PSK): a digital modulation technique for wireless data transmission from a biomedical device. PSK modulates the phase of the carrier signal according to the digital input: in the simplest scheme a zero input leaves the phase of the carrier signal the same, whereas a one introduces a 180° phase shift.

Photodiode: a device that transduces an optical signal into an electrical one using the photovoltaic effect, i.e. the generation of a voltage across a positive-negative (P-N) junction of a semiconductor when the junction is exposed to light.

Piezoelectric sensor: a sensor that converts changes in pressure into an electric signal.

Pinna: the visible part of the ear that extends outside the head. The pinna collects sound and acts as a "funnel" to amplify and direct it towards the auditory canal.

Postsynaptic potential: a change in the membrane potential of the postsynaptic terminal of a chemical synapse. Postsynaptic potentials initiate or inhibit action potentials.

Pre-market approval: the Food and Drug Administration process of scientific and regulatory review to evaluate the safety and effectiveness of Class III medical devices.

Precision: the precision of a device is defined as the smallest difference between separate measurements that can be detected.

Pulse oximeter: a device used to measure blood oxygenation non-invasively. Pulse oximeters are small, inexpensive devices that can be placed around a finger, toe or ear lobe, and can be battery operated for home-monitoring.

QRS-complex: a feature in the ECG signal which is produced by the depolarization of the right and left ventricles. Within the QRS-complex the Q-wave is generated when the depolarization wave travels through the interventricular septum, the R-wave corresponds to a contraction of the left ventricle to pump blood out of the ventricle, and the S-wave to the basal sections of the ventricles depolarizing due to the contraction of the right ventricle.

Quantization noise: the quantization noise of an analogue-to-digital converter is the difference between the analogue input signal and the digitized output signal. This error becomes smaller the higher the resolution of the analogue-to-digital converter.

Receiver operator characteristic (ROC): for a given clinical diagnosis the ROC is a graph that shows the true positive rate on the vertical axis versus the false positive rate on the horizontal axis. The ROC can be used to assess the sensitivity and specificity of a diagnosis which is based on measurements from a particular medical device.

Reduced amplitude zone: a feature in high frequency electrocardiography analysis defined when at least two local maxima or two local minima are present within one high-frequency QRS envelope, and when each local maximum or minimum is defined by its having an absolute voltage that is higher or lower than the three envelope sample points immediately preceding and following it.

Refractory period: during membrane depolarization and repolarization there are two different types of refractory period, absolute and relative. During the absolute refractory period, if the cell receives another stimulus, no matter how large this stimulus it does not lead to a second action potential. During the relative refractory period, a higher-than-normal stimulus is needed in order to produce an action potential.

Release/disengaging time: the release/disengaging time in a digital hearing aid compressor circuit is defined as the time required for the signal to reach within 2 dB of its final value after the input sound pressure level decreases from 80 to 55 dB.

Reproducibility: a measure of the difference between successive measurements taken under identical measuring conditions.

Residual current device (RCD): an electrical circuit used in patient treatment areas in hospitals to sense any leakage currents flowing to earth from medical equipment. If a significant leakage current (10 mA) is detected, the RCD shuts off the power supplied to the equipment within ~40 ms.

Resonant circuits: electrical circuits which have a particular frequency, the resonant frequency, at which the output is at maximum. The impedance of the circuit is purely resistive at this frequency, i.e. the imaginary component is zero.

Sallen-Key filters: the most common topology for an active filter. They can be designed to have the characteristics of a Butterworth, Chebyshev, Bessel, elliptical or any other filter.

Sample-and-hold (S/H) circuit: an electronic circuit which samples an analogue signal and holds this sample while the rest of the circuitry in the analogue-to-digital converter derives the equivalent digital output.

Sensitivity: the sensitivity of a device is defined as the number of true positives divided by the sum of the true positives and the false negatives.

Sensorineural hearing loss: refers to two related mechanisms which together cause hearing loss, namely sensory loss involving the inner ear and neural loss involving the hearing nerve. Sensorineural hearing loss can be treated either by hearing aids that magnify the sound signals, or cochlear implants that stimulate the nerve cells directly.

Shift register: an electronic device that produces a discrete delay of a digital signal or waveform. An n-stage shift register delays input data by n discrete cycles of an internal clock.

Shot noise: one component of noise in biomedical devices, shot noise is caused by the random arrival time of electrons (or photons), and can be modelled as a noise current.

Signal averaging: a postprocessing technique in which successively-acquired signals are added together to increase the signal-to-noise. If the noise is random (Gaussian) then the signal-to-noise is proportional to the square root of the number of co-added scans.

Slow cortical potential (SCP): very low frequency (<1 Hz) changes in the electro-encephalography signals. A negative SCP is produced by increased, and a positive SCP by lowered, neuronal activity.

Smart textiles: materials used in clothing which incorporate sensors to measure different physiological parameters such as temperature, an electrocardiogram, and skin resistance.

Sodium-potassium pump: an enzyme-based "molecular pump" that maintains cellular concentrations of sodium and potassium ions by transporting sodium ions from inside the cell to outside the cell, and potassium ions from outside the cell to the inside.

Sound pressure level: defined as the amount in dB that a sound pressure wave exceeds the reference value of 2×10^{-5} Pascals.

Specific absorption rate: the amount of power deposited in a patient by a medical device, measured in Watts/kg.

Specificity: the specificity of the device is the number of true negatives divided by the sum of the number of true negatives and false positives.

Stochastic signal: a type of signal for which future values cannot be predicted based on past ones (in contrast to deterministic signals). However, certain properties such as the standard deviation of stochastic signals may be time-independent, and so can be predicted.

Strain gauge: a sensor whose resistance depends upon the strain imposed upon it. One application of a strain gauge in biomedical devices is monitoring changes in corneal curvature which occur in response to variations in intraocular pressure in patients with glaucoma.

T-wave: a feature of the ECG signal produced by ventricular repolarization.

Tachycardia: a faster-than-normal heartbeat. There are many non-specific causes including damage to heart tissues, congenital abnormality of the heart or an overactive thyroid.

Telecoil: an inductively coupled loop present in a digital hearing aid, which is used to receive an externally transmitted signal. Public places which transmit signals typically have a sign saying "Hearing loop installed: switch hearing aid to T-coil". Using a telecoil increases the signal-to-noise and reduces the degree of room reverberation effects.

Theta waves: components of the electroencephalography signal in the 4 to 7 Hz range. Low amplitude theta waves are detected in a normal awake adult, in contrast to the larger theta waves seen in children, as well as adults in light sleep states. As is the case for delta waves, a large amount of theta activity in awake adults is related to various neurological diseases.

Transducer: a device that transforms one form of energy into another. Transducers used in biomedical devices convert, for example, light/pressure/motion/sound into either voltages or currents.

Transfer function: a mathematical expression which represents the frequency-dependence of the output of an electrical circuit as a function of its input.

Transimpedance amplifier: an electrical circuit which converts current to voltage, most often implemented using an operational amplifier. A transimpedance amplifier is used with a sensor which has a current response that is more linear than its voltage response, for example a photodiode used in a pulse oximeter.

Type 1 diabetes: a form of diabetes mellitus in which insufficient insulin is produced, resulting in high blood sugar levels in the body. The underlying mechanism involves autoimmune destruction of the pancreatic insulin-producing beta cells.

Type 2 diabetes: the most common form of diabetes mellitus, representing a long-term metabolic disorder characterized by high blood sugar, insulin resistance, and low insulin levels. Type 2 diabetes primarily occurs as a result of obesity and lack of exercise, with an increasing proportion of the population becoming genetically more at risk.

Value-based medicine: a concept which represents the trade-off between improved healthcare *versus* potential increased costs to the healthcare system. The "value" of a new drug or device is calculated, with its value defined as the degree of improvement relative to its increased cost (which also includes associated costs in terms of parameters such as the need for highly trained operators and additional training).

Ventricular fibrillation: one of the most serious disturbances of the cardiac rhythm disturbance, in which the heart's electrical activity becomes disordered, causing the ventricles to contract in a rapid, unsynchronized way. The result is that very little blood is pumped out of the heart.

Visual evoked potential (VEP): a change in the electroencephalography signals which occur in the visual cortex when there is a visual stimulation. The magnitude of the VEP depends upon the position of the stimulus within the visual field, with the maximum response corresponding to the centre of the visual field.

Volume conductor: refers to a body part through which ionic currents propagate to the body surface where they are indirectly detected using surface electrodes. A volume conductor can be modelled very simply by assuming a single homogeneous tissue type, or much more realistic models can be used with accurate geometries of each of the different organs and tissues with appropriate electrical properties.

Wheatstone bridge: an electrical circuit which usually consists of four lumped elements (typically resistors or capacitors), three of which have fixed values and the remaining one being the particular sensor whose impedance depends upon the physical parameter being measured. Changes in the output voltage of the circuit can be directly related to changes in the impedance of the sensor.

ZigBee: a low-power data transmission protocol which operates in the 906–924 MHz and the 2405–2480 MHz ranges, and is used in some implanted wireless devices.

Index

Printed in the United States
By Bookmasters